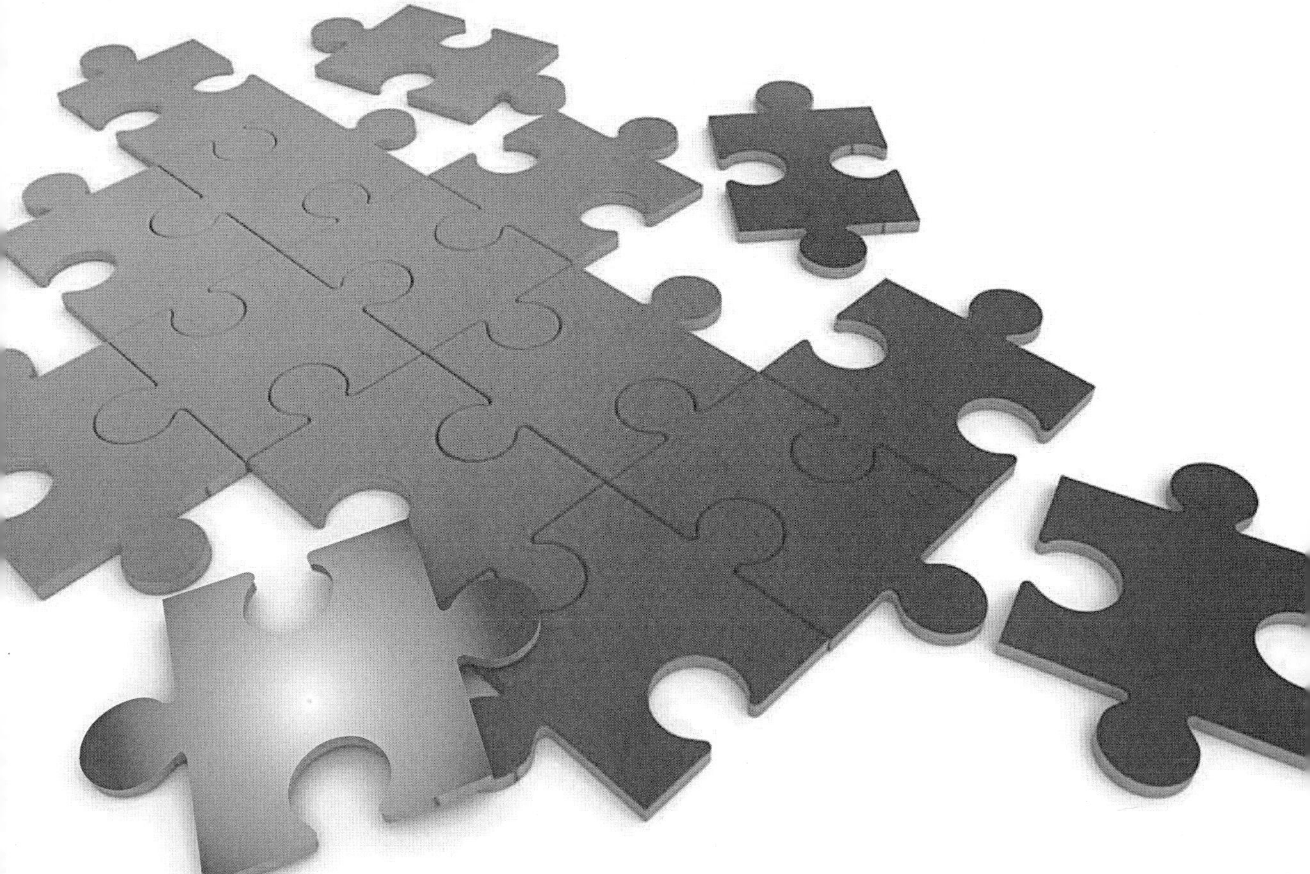

다중물리 해석학 입문

(Introduction to Analysis of Multiphysics)

서 용 권 · 강 상 모 공 저

홍릉과학출판사

머리말

다중 물리학(multi-physics)은 물리학의 다양한 학문 분야를 동시에 사용하여 자연 현상과 그 이치를 분석하고 응용하기 위한 학문이다. 이 용어는 비단 물리학 내의 세부 학문분야끼리의 연계뿐만 아니라 넓게는 화학이나 생물학 혹은 생명공학과 같은 다른 학문 분야가 어우러져 혼합 활용되는 다중 학문(multi-disciplinary)을 포함하여 일컫는 경우가 많다.

최근 다중 물리학에 대한 연구에 관심이 집중되고 있는 것은 어쩌면 당연한 귀결인지도 모른다. 모든 자연 현상과 공학적 문제는 다양한 학문적 원리가 결합된 형태로 나타나고 또 그러한 통합적 개념으로 이해해야 하기 때문이다. 지금까지는 그러한 필요성을 인식하였다 하더라도 문제를 통합적으로 풀 수 있는 도구가 없었으나, 각 학문 분야별로 컴퓨터를 이용한 계산 및 설계 도구들이 많은 발전을 해 왔고 최근에 와서는 이들 도구들이 결합된 형태로 제공되기 때문에 이제는 본격적인 다중 물리학의 연구가 가능하게 된 것이다.

다중 물리학을 대학원 과정에서 가르치기 위한 교재는 아직은 찾기 어렵다. 그 이유로는 관련되는 다양한 학문 분야를 제대로 이해하는 것이 현실적으로 어렵고 또 일반성을 해치지 않고 동시에 이해하기 쉬운 용어로 관련 현상을 설명하고 문제 해결을 제시하는 일이 쉽지 않기 때문이라 생각된다.

본 교재는 그럼에도 불구하고 2011년 한국에너지기술평가원의 인력양성사업의 지원에 힘입어 기계공학에서의 유체역학, 전기공학에서의 정전기학, 화학공학에서의 전기화학 분야를 통합하는 교재개발에 힘 쓴 결과이다. 비록 교재 제목은 다중 물리학이지만, 원고를 완성하고 보니 솔직히 관련 학문들의 내용을 간추려서 병렬적으로 나열한 느낌이 없지 않다. 이들의 진면목한 결합을 위해서는 향후에 다중 물리학 전문 소프트웨어(예를 들면 COMSOL)를 활용한 예들을 많이 추가해야 할 것으로 생각된다.

교재에 오타나 오류가 있다면 이는 다분히 저자들의 책임이며 그로 인해 발생되는 그 어떠한 불편함이나 불이익에 대해서는 도덕적 책임감을 느낀다. 이 교재의 출판 이후 각종 오타나 오류에 대한 독자들의 피드백이 절실히 요청된다.

본 교재의 저술에 있어서 주로 참고한 문헌은 다음 쪽에 나열한 바와 같다.

끝으로, 산업통상자원부 및 한국에너지기술평가원(에너지인력양성사업)의 관계자, 원고 작성 중 격려를 아끼지 않은 사업단의 참여 교수들, 그리고 출판을 허락해 준 홍릉과학출판사 관계자에게 감사드린다.

2013년 6월 저자 일동

List of References

Chapter 1:

M. H. Sadd, "Elasticity: Theory, Applications, and Numerics", Elsevier, 2005.
A. P. S. Selvadurai, "Partial Differential Equations in Mechanics 2", Springer, 2000.

Chapter 2, 3:

I. G. Currie, "Fundamental Mechanics of Fluids", McGraw-Hill Book Co., 1974.

Chapter 4:

F. P. Incropera and D. P. DeWitt, "Fundamentals of Heat and Mass Transfer", John Wiley and Sons, Inc., 2002.

Chapter 5:

J. E. Brady, J. W. Russel and J. R. Holum, "Chemistry", John Wiley and Sons, Inc., 2000.

Chapter 6:

K. J. Laidler and J. H. Meiser, "Physical Chemistry", Houghton Mifflin Co., 1999.

Chapter 7:

D. Halliday, R. Resnick and J. Walker, "Fundamentals of Physics", John Wiley and Sons, Inc., 1997.
R. Wangness, "Electromagnetic Fields", John Wiley and Sons, Inc., 1986.
E. M. Purcell, "Electricity and Magnetism", McGraw-Hill Book Co., 1985.

차례

Chapter 1 Elasticity · 1

Chapter 2 Laminar Flow: Governing Equations · 41

Chapter 3 Laminar Flow: Exact Solutions · 59

Chapter 4 Heat Transfer · 87

Chapter 1. Elasticity

1.1 Index Notation and Transformation

1.1.1 Index notation

○ Free index

e.g., vector; $a_i = \begin{bmatrix} a_1 \\ a_2 \\ a_3 \end{bmatrix}$, tensor; $a_{ij} = \begin{bmatrix} a_{11} & a_{12} & a_{13} \\ a_{21} & a_{22} & a_{23} \\ a_{31} & a_{32} & a_{33} \end{bmatrix}$

$$\boldsymbol{c} = \boldsymbol{a} + \boldsymbol{b} \Rightarrow c_i = a_i + b_i \qquad (i = 1,2,3;\ \text{free index})$$

○ Repeated index (Einstein summation)

e.g., $a_i b_i = a_1 b_1 + a_2 b_2 + a_3 b_3$

$c_{kk} = c_{11} + c_{22} + c_{33}$

$a_{pq} b_q = a_{p1} b_1 + a_{p2} b_2 + a_{p3} b_3$ (p is free and q is repeated)

$$a_{ij} x_i x_j = (a_{11} x_1 x_1 + a_{12} x_1 x_2 + a_{13} x_1 x_3) + (a_{21} x_2 x_1 + a_{22} x_2 x_2 + a_{23} x_2 x_3) + (a_{31} x_3 x_1 + a_{32} x_3 x_2 + a_{33} x_3 x_3)$$

ex) $\mathrm{x}' = \mathrm{Ax}$

$$\Rightarrow \begin{cases} x_1{}' = a_{11} x_1 + a_{12} x_2 + a_{13} x_3 \\ x_2{}' = a_{21} x_1 + a_{22} x_2 + a_{23} x_3 \\ x_3{}' = a_{31} x_1 + a_{32} x_2 + a_{33} x_3 \end{cases}$$

$$\Rightarrow \begin{cases} x_1{}' = a_{1m} x_m \\ x_2{}' = a_{2m} x_m \\ x_3{}' = a_{3m} x_m \end{cases}$$

$$\Rightarrow x_i{}' = a_{im} x_m$$

m: repeated (summation)

i: not repeated (free index)

ex) $T_{ij} = A_{im} A_{jm}$

$m = 1,2,3$: repeated (summation)

$i, j = 1,2,3$: not repeated (free indices)

$$\begin{cases} T_{11} = A_{11} A_{11} + A_{12} A_{12} + A_{13} A_{13} \\ T_{12} = A_{11} A_{21} + A_{12} A_{22} + A_{13} A_{23} \\ T_{13} = A_{11} A_{31} + A_{12} A_{32} + A_{13} A_{33} \\ T_{21} = A_{21} A_{11} + A_{22} A_{12} + A_{23} A_{13} \\ T_{22} = \cdots\cdots\cdots \\ \cdots\cdots\cdots\cdots \end{cases} \Rightarrow 9 \text{ equations}$$

○ Kronecker delta; $\delta_{ij} = \begin{cases} 1 & \text{if } i = j \\ 0 & \text{if } i \neq j \end{cases}$

e.g., $\delta_{ii} = 3, \quad \delta_{ij} a_j = a_i, \quad \delta_{pq} a_{qk} = a_{pk}$

ex) $\delta_{11} = \delta_{22} = \delta_{33} = 1$

$\delta_{12} = \delta_{21} = \delta_{13} = \ldots = 0$

ex) $I = [\delta_{ij}]$ (unit matrix)

$\delta_{ii} = 3$

$\delta_{im} a_m = a_i$

$\delta_{im} T_{mj} = T_{ij}$

ex) If e_1, e_2 and e_3 are orthogonal unit vectors, $e_i \cdot e_j = \delta_{ij}$

○ Alternating (or permutation) symbol; $\epsilon_{ijk} = \begin{cases} +1 & \text{if } ijk \text{ is an even permutation} \\ -1 & \text{if } ijk \text{ is an odd permutation} \\ 0 & \text{otherwise} \end{cases}$

e.g., $\epsilon_{123} = 1, \ \epsilon_{232} = 0, \ \epsilon_{231} = 1, \ \epsilon_{132} = -1, \ \epsilon_{ijk}\delta_{jk} = 0$

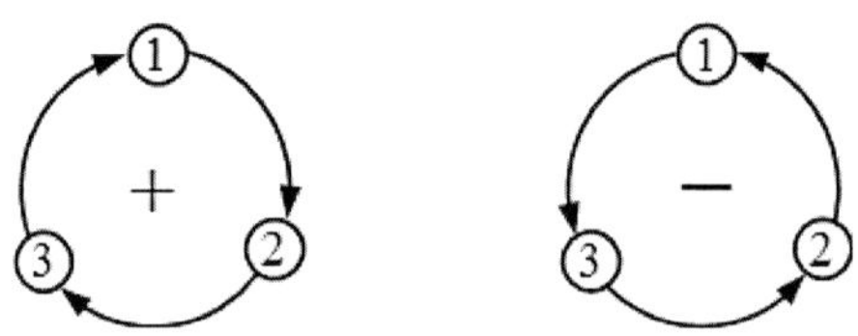

Fig. 1.1 Permutation of three consecutive numbers

ex) $\varepsilon_{123} = \varepsilon_{231} = \varepsilon_{312} = 1$

$\varepsilon_{321} = \varepsilon_{213} = \varepsilon_{132} = -1$

$\varepsilon_{113} = \varepsilon_{122} = \varepsilon_{313} = \ldots = 0$

ex) $e_i \times e_j = \varepsilon_{ijk} e_k$

$e_1 \times e_2 = e_3$

$e_2 \times e_1 = - e_3$

$e_1 \times e_1 = 0$

....

ex) $a = a_i e_i$ and $b = b_j e_j$

$a \times b = (a_i e_i) \times (b_j e_j) = \varepsilon_{ijk} a_i b_j e_k$

○ Manipulation

- Substitution

ex) $a_i = U_{im} b_m$

$b_i = V_{im} c_m$

To substitute the second equation into the first equation, we first change the free index i of the second equation to m and simultaneously the dummy index m to another letter, say, n:

$b_m = V_{mn} c_n \Rightarrow a_i = U_{im} b_m = U_{im} V_{mn} c_n$

- Multiplication

ex) $\begin{cases} p = a_m b_m \\ q = c_m d_m \end{cases} \Rightarrow pq = a_m b_m c_n d_n$

1.1.2 Coordinate transformation

We need rules of transformation of vectors and tensors between different coordinate systems in order to derive important relations among the quantities such as stress, strain, and displacement. Suppose two different Cartesian coordinates (x_1, x_2, x_3) and (x'_1, x'_2, x'_3) differ only by orientation (Fig. 1.2). The unit vectors in each coordinate system are defined as $\{e_1, e_2, e_3\}$ and $\{e'_1, e'_2, e'_3\}$. Let Q_{ij} denote the cosine of the angle between the x'_i-axis and the x_j-axis, $Q_{ij} = \cos(x'_i, x_j)$: Note that $Q_{ij} \neq Q_{ji}$. Then, for instance, the unit vector e'_1 can be written as the sum, $e'_1 = Q_{11} e_1 + Q_{12} e_2 + Q_{13} e_3$. In general, we can write

$$e'_i = Q_{ij} e_j$$

or inversely

$$e_i = Q_{ji} e'_j$$

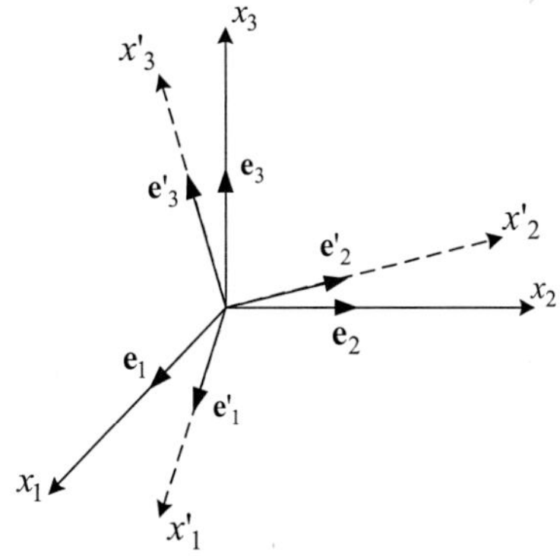

Fig. 1.2 A new Cartesian coordinate system (x'_1, x'_2, x'_3) constructed by rotation of the existing one (x_1, x_2, x_3).

Any arbitrary vector v can be written in terms of each set of the unit vectors for the two coordinate systems like $v = v_i e_i = v'_i e'_i$. Substituting above two formula relating e_i and e'_i into this, we obtain

$$v'_i = Q_{ij} v_j$$

or

$$v_i = Q_{ji} v'_j$$

From these results we derive $v_i = Q_{ji} Q_{jk} v_k$. The left-hand side is equivalently written as $\delta_{ik} v_k$ implying that

$$Q_{ji} Q_{jk} = \delta_{ik}$$

which corresponds to the orthogonal condition.

In a similar fashion, any arbitrary second-order tensor $\boldsymbol{A}$ (its component being a_{ij}) can be written by using dyadic notation as follows.

$$\boldsymbol{A} = a_{ij} e_i e_j = a'_{ij} e'_i e'_j$$

Note that the indices i and j on both sides are repeated and thus we can replace i and j on the left-hand side by p and q respectively: $a_{pq} e_p e_q = a'_{ij} e'_i e'_j$. Substituting for e_p and e_q provides $e_p e_q = Q_{ip} e'_i Q_{jq} e'_j = Q_{ip} Q_{jq} e'_i e'_j$. Then we arrive at the transformation rule for the second-order tensor:

$$a'_{ij} = Q_{ip} Q_{jq} a_{pq}$$

which can also be written as

$$\boldsymbol{A}' = (\boldsymbol{QA}) \boldsymbol{Q}^T$$

Homework

A certain vector and a tensor are given in a particular coordinate frame as

$$a_i = \begin{bmatrix} 1 \\ 2 \\ 3 \end{bmatrix}, \quad b_{ij} = \begin{bmatrix} 1 & 0 & -1 \\ 0 & 2 & 2 \\ 3 & 0 & 1 \end{bmatrix}$$

A new frame of coordinate is created by rotating the coordinates 60° about the axis x_3 in the counter-clockwise direction. Determine the matrix Q_{ij}, and derive three components of a'_i and nine components of b'_{ij}.

1.2 Displacement, Deformation and Strains

1.2.1 Displacement

Suppose a vertical bar's bottom surface is attached to a horizontal wall and is subjected to a shear force on the top surface as shown in Fig. 1.3. A material point initially at P will move to a new point Q after the force has been applied. The displacement vector at the point P is then defined as $\mathrm{u} = \overrightarrow{PQ}$.

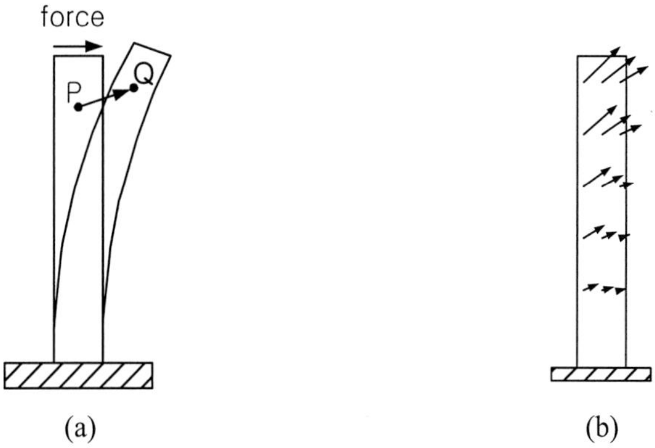

Fig. 1.3 A vertical bar attached to a bottom wall subjected to shear force on the top surface: (a) displacement of a point P under forcing; (b) the resultant displacement vector field.

1.2.2 Deformation and strain

The displacement vector u introduced above only specifies the displaced position of a specific point in the material under investigation. Stress, which is outcome of external force acting on the material, is however related to strain, which is concerned with deformation of infinitesimal material element, not to the displacement itself. Thus, it is possible that even though there may be displacement, the strain can be zero (Fig. 1.4).

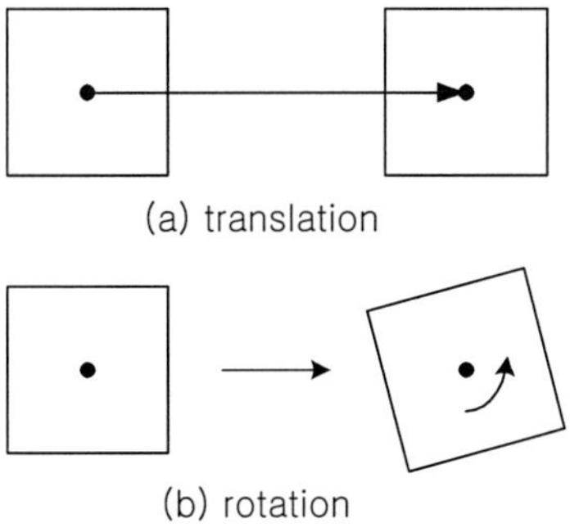

Fig. 1.4 Two kinds of displacements showing no deformation (and so, no strain): (a) pure translation; (b) pure rotation.

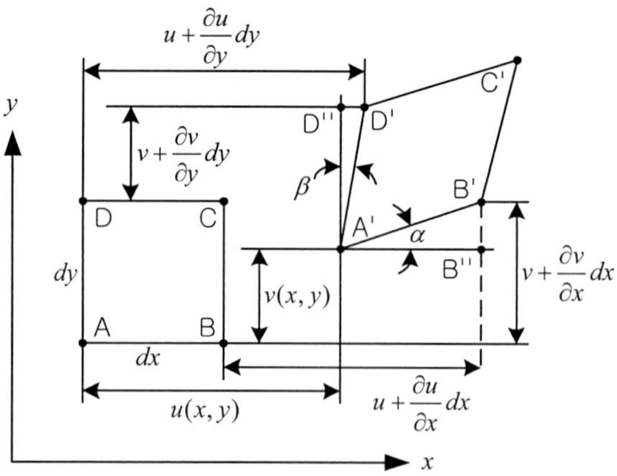

Fig. 1.5 The initial material element ABCD and the element A'B'C'D' displaced as well as deformed under a small forcing.

Consider a material element of rectangular shape, ABCD, as shown in Fig. 1.5. Displacement vectors of points A, B and D are

point A → A'; (u,v)

point B → B'; $(u+\frac{\partial u}{\partial x}dx, v+\frac{\partial v}{\partial x}dx)$

point D → D'; $(u+\frac{\partial u}{\partial y}dy, v+\frac{\partial v}{\partial y}dy)$

The normal strain e_x defined as $e_x = \frac{A'B' - AB}{AB}$ is

$$e_x = \frac{\partial u}{\partial x}$$

In a similar fashion, the normal strain $e_y = \frac{A'D' - AD}{AD}$ is

$$e_y = \frac{\partial v}{\partial y}$$

The shear strain defined as $e_{xy} = (\alpha+\beta)/2$ is

$$e_{xy} = \frac{1}{2}\left(\frac{B'B''}{AB}+\frac{D'D''}{AD}\right)=\frac{1}{2}\left(\frac{\partial u}{\partial y}+\frac{\partial v}{\partial x}\right)$$

The formulation presented so far was for the 2-dimensional case. In general for the 3-dimensional case, the strain tensor can be written in terms of the displacement vector components as follows.

$$e_{ij} = \frac{1}{2}\left(\frac{\partial u_i}{\partial x_j}+\frac{\partial u_j}{\partial x_i}\right) \tag{1.1a}$$

For later use, we also define the rotation tensor ω_{ij} as follows.

$$\omega_{ij} = \frac{1}{2}\left(\frac{\partial u_i}{\partial x_j} - \frac{\partial u_j}{\partial x_i}\right) \tag{1.1b}$$

Homework

For the displacement vector $\mathrm{u} = (x^2y,\ 2y^2,\ xz^2)$, derive all the components of the strain tensor e_{ij}.

1.2.3 Strain transfomation

Since the strain constitutes a second-order tensor, we can apply the transformation rule. The strain e_{ij} in a coordinate system (x_1, x_2, x_3) is transformed to e'_{ij} in a different coordinate system (x'_1, x'_2, x'_3), where

$$e'_{ij} = Q_{ip}Q_{jq}e_{pq}$$

For a 2-dimensional case, where the coordinate system for (x'_1, x'_2, x'_3) is constructed by rotating the frame about the x_3-axis by the angle θ, Q_{ij} is given as

$$Q_{ij} = \begin{bmatrix} \cos\theta & \sin\theta & 0 \\ -\sin\theta & \cos\theta & 0 \\ 0 & 0 & 1 \end{bmatrix}$$

Then,

$$\begin{aligned} e'_x &= e'_{11} = Q_{1p}Q_{1q}e_{pq} = Q_{11}Q_{11}e_{11} + Q_{11}Q_{12}e_{12} + Q_{12}Q_{11}e_{21} + Q_{12}Q_{12}e_{22} \\ &= e_x\cos^2\theta + 2e_{xy}\sin\theta\cos\theta + e_y\sin^2\theta \\ &= \frac{e_x + e_y}{2} + \frac{e_x - e_y}{2}\cos2\theta + e_{xy}\sin2\theta \end{aligned}$$

Similarly,

$$\begin{aligned} e'_y &= e'_{22} = e_y\cos^2\theta - 2e_{xy}\sin\theta\cos\theta + e_x\sin^2\theta \\ &= \frac{e_x + e_y}{2} - \frac{e_x - e_y}{2}\cos2\theta - e_{xy}\sin2\theta \end{aligned}$$

$$\begin{aligned} e'_{xy} &= e'_{12} = -e_x\sin\theta\cos\theta + e_{xy}(\cos^2\theta - \sin^2\theta) + e_y\sin\theta\cos\theta \\ &= -\frac{e_x - e_y}{2}\sin2\theta + e_{xy}\cos2\theta \end{aligned}$$

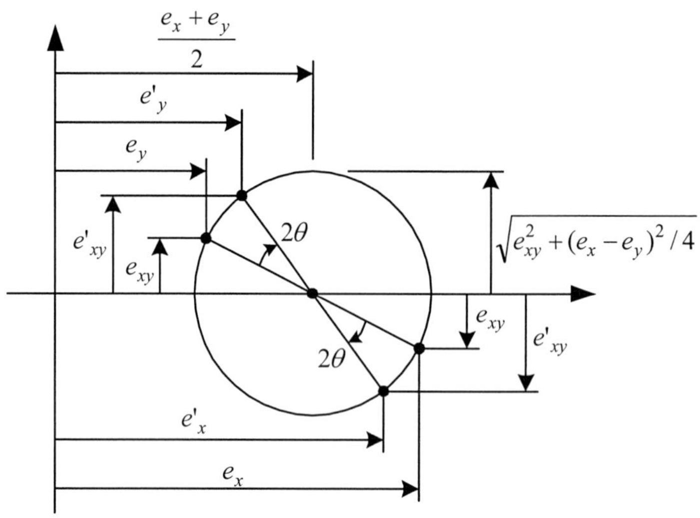

Fig. 1.6 Mohr circle.

For a 2-dimensional case, we can construct the Mohr circle (Fig. 1.6) to understand the relation between the strains in the original coordinates, e_x, e_y, e_{xy} and those in the new coordinates, e'_x, e'_y, e'_{xy}. It is seen that the maximum normal strain is $e_m = \bar{e}+\tau_m$ and the maximum shear strain is τ_m, where $\bar{e}=(e_x+e_y)/2$ and $\tau_m = \sqrt{e_{xy}^2+(e_x-e_y)^2/4}$ are invariant of the coordinates employed; i.e.,

$$\bar{e}=(e_x+e_y)/2=(e'_x+e'_y)/2$$

$$\tau_m = \sqrt{e_{xy}^2+(e_x-e_y)^2/4}=\sqrt{e'^2_{xy}+(e'_x-e'_y)^2/4}$$

Homework

Show that the number pair (e'_x, e'_{xy}) or (e'_y, e'_{xy}) describes the Mohr circle.

1.2.4 Strain compatibility

It was shown previously, i.e., Eq. (1.1a), that the strain tensor can be represented by the spatial gradient of the displacement vector. On the other hand, in a 3-dimensional case, the number of components of the strain tensor is 6 whereas that of the displacement vector is 3. This implies that three of the 6 components in the strain tensor can be written in terms of the other 3, which serves as the underlying concept of the compatability condition in solving elasticity problems.

In the 2-dimensional case, for example, we have

$$e_x = \frac{\partial u}{\partial x}, \quad e_y = \frac{\partial v}{\partial y}, \quad e_{xy} = \frac{1}{2}\left(\frac{\partial u}{\partial y} + \frac{\partial v}{\partial x}\right)$$

and it is seen that two displacement components u and v are used to represent three strain components e_x, e_y and e_{xy}; so, it is possible to eliminate u and v to derive the relation among e_x, e_y and e_{xy}. Taking derivative of e_{xy} with respect to x and y yields

$$\frac{\partial^2 e_{xy}}{\partial x \partial y} = \frac{1}{2}\left(\frac{\partial^3 u}{\partial x \partial y^2} + \frac{\partial^3 v}{\partial y \partial x^2}\right)$$

which gives

$$\frac{\partial^2 e_x}{\partial y^2} + \frac{\partial^2 e_y}{\partial x^2} = 2\frac{\partial^2 e_{xy}}{\partial x \partial y}$$

This corresponds to a compatibility condition for the 2-dimensional problem.

In this way, we can eliminate the displacement components for the 3-dimensional case to get the followings.

$$\frac{\partial^2 e_x}{\partial y^2} + \frac{\partial^2 e_y}{\partial x^2} = 2\frac{\partial^2 e_{xy}}{\partial x \partial y} \tag{1.2a}$$

$$\frac{\partial^2 e_y}{\partial z^2} + \frac{\partial^2 e_z}{\partial y^2} = 2\frac{\partial^2 e_{yz}}{\partial y \partial z} \tag{1.2b}$$

$$\frac{\partial^2 e_z}{\partial x^2} + \frac{\partial^2 e_x}{\partial z^2} = 2\frac{\partial^2 e_{zx}}{\partial z \partial x} \tag{1.2c}$$

$$\frac{\partial^2 e_x}{\partial y \partial z} = \frac{\partial}{\partial x}\left(-\frac{\partial e_{yz}}{\partial x} + \frac{\partial e_{zx}}{\partial y} + \frac{\partial e_{xy}}{\partial z}\right) \tag{1.2d}$$

$$\frac{\partial^2 e_y}{\partial z \partial x} = \frac{\partial}{\partial y}\left(-\frac{\partial e_{zx}}{\partial y} + \frac{\partial e_{xy}}{\partial z} + \frac{\partial e_{yz}}{\partial x}\right) \tag{1.2e}$$

$$\frac{\partial^2 e_z}{\partial x \partial y} = \frac{\partial}{\partial z}\left(-\frac{\partial e_{xy}}{\partial z} + \frac{\partial e_{yz}}{\partial x} + \frac{\partial e_{zx}}{\partial y}\right) \tag{1.2f}$$

These are called *Saint-Venant compatability equations*. However, these six equations are not all independent. Only three should be independent by the reason discussed previously.

1.2.5 Cylindrical and spherical coordinates

In cylindrical coordinates (r, θ, z) (Fig. 1.7), the relation between the strain tensor and the displacement vector (u_r, u_θ, u_z) is;

$$e_r = \frac{\partial u_r}{\partial r}, \; e_\theta = \frac{1}{r}\left(u_r + \frac{\partial u_\theta}{\partial \theta}\right), \; e_z = \frac{\partial u_z}{\partial z}$$

$$e_{r\theta} = \frac{1}{2}\left(\frac{1}{r}\frac{\partial u_r}{\partial \theta} + \frac{\partial u_\theta}{\partial r} - \frac{u_\theta}{r}\right)$$

$$e_{z\theta} = \frac{1}{2}\left(\frac{1}{r}\frac{\partial u_z}{\partial \theta} + \frac{\partial u_\theta}{\partial z}\right)$$

$$e_{zr} = \frac{1}{2}\left(\frac{\partial u_r}{\partial z} + \frac{\partial u_z}{\partial r}\right)$$

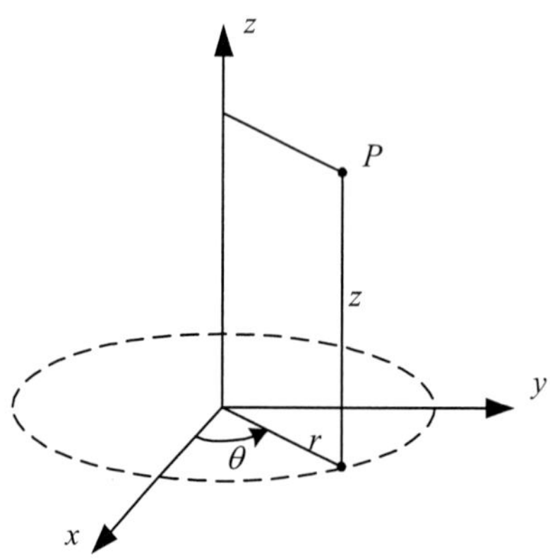

Fig. 1.7 Cylindrical coordinates.

In spherical coordinates (R,ϕ,θ) (Fig. 1.8), the relation between the strain tensor and the displacement vector (u_R, u_ϕ, u_θ) is;

$$e_R = \frac{\partial u_R}{\partial R}$$

$$e_\phi = \frac{1}{R}\left(u_R + \frac{\partial u_\phi}{\partial \phi}\right)$$

$$e_\theta = \frac{1}{R\sin\phi}\left(\frac{\partial u_\theta}{\partial \theta} + u_R\sin\phi + u_\phi\cos\phi\right)$$

$$e_{R\phi} = \frac{1}{2}\left(\frac{1}{R}\frac{\partial u_R}{\partial \phi} + \frac{\partial u_\phi}{\partial R} - \frac{u_\phi}{R}\right)$$

$$e_{\phi\theta} = \frac{1}{2R}\left(\frac{1}{\sin\phi}\frac{\partial u_\phi}{\partial \theta} + \frac{\partial u_\theta}{\partial \phi} - u_\theta\cot\phi\right)$$

$$e_{R\theta} = \frac{1}{2}\left(\frac{1}{R\sin\phi}\frac{\partial u_R}{\partial \theta} + \frac{\partial u_\theta}{\partial R} - \frac{u_\theta}{R}\right)$$

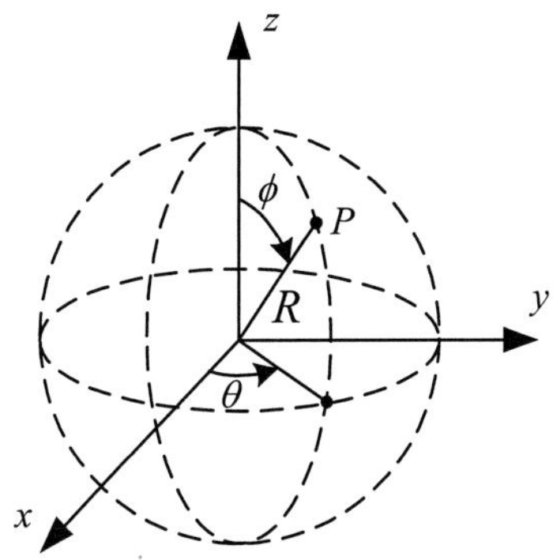

Fig. 1.8 Spherical coordinates.

1.3 Stress and Equilibrium

1.3.1 Body force and surface force externally applied

○ Body force; force acting on the volume of material of interest, proportional to the volume

Body force density; body force per unit volume

○ Surface force; force acting on the surface of material of interest, proportional to the surface area (Fig. 1.9)

Surface traction; surface force per unit area

1.3.2 Normal and shear stress

○ Normal stress; the normal component of the surface force per unit area acting on the surface of interest

Shear stress; the tangential component of the surface force per unit area acting on the surface of interest

To define a stress, we need to specify the orientation of the surface normal as well as its direction of action. Thus, the stress must be of a second-order tensor. Thus, σ_{ij}, for instance, denotes the stress acting on the i-plane along the x_j-direction; e.g., σ_{12} (or τ_{xy}) is the stress acting on the x-plane along the y-direction.

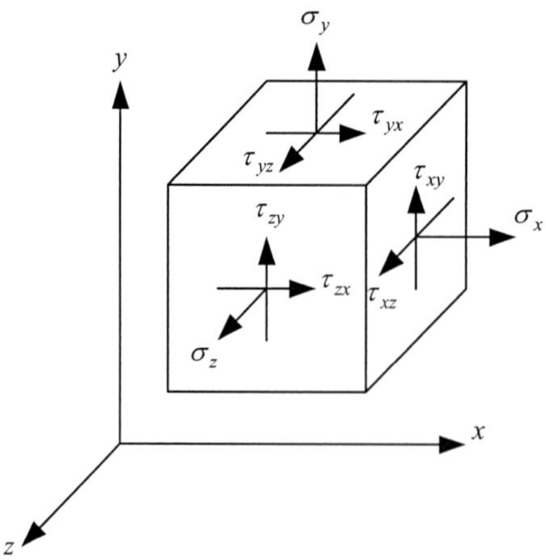

Fig. 1.9 Normal (σ) and tangential (τ) stresses acting on the three frontal surfaces of an infinitesimal 3-D material element.

1.3.3 Stress transformation

Similar to the strain, transformation of the stress can be written as

$$\sigma'_{ij} = Q_{ip}Q_{jq}\sigma_{pq}$$

For a 2-dimensional case with

$$Q_{ij} = \begin{bmatrix} \cos\theta & \sin\theta & 0 \\ -\sin\theta & \cos\theta & 0 \\ 0 & 0 & 1 \end{bmatrix}$$

we have

$$\sigma'_x = \frac{\sigma_x + \sigma_y}{2} + \frac{\sigma_x - \sigma_y}{2}\cos 2\theta + \tau_{xy}\sin 2\theta$$

$$\sigma'_y = \frac{\sigma_x + \sigma_y}{2} - \frac{\sigma_x - \sigma_y}{2}\cos 2\theta - \tau_{xy}\sin 2\theta$$

$$\sigma'_{xy} = -\frac{\sigma_x - \sigma_y}{2}\sin 2\theta + \tau_{xy}\cos 2\theta$$

We can see that the transformation takes the same form as that for the strain. So, we can understand the relation between σ_{ij} and σ'_{ij} by using the same Mohr circle. We also note that sum of the normal stress components (trace of the stress tensor), $\sigma_x + \sigma_y$, is invariant of the coordinate system. In fact, this also applies to a 3-dimensional case; $\sigma'_x + \sigma'_y + \sigma'_z = \sigma_x + \sigma_y + \sigma_z$.

We can also use the above formula in deriving the radial and azimuthal components of the stress in the polar coordinate system from those of the Cartesian coordinate system; that is, we can use the identity, $\sigma_r = \sigma'_x$, $\sigma_\theta = \sigma'_y$ and $\tau_{r\theta} = \sigma'_{xy}$ as shown in

Fig. 1.10.

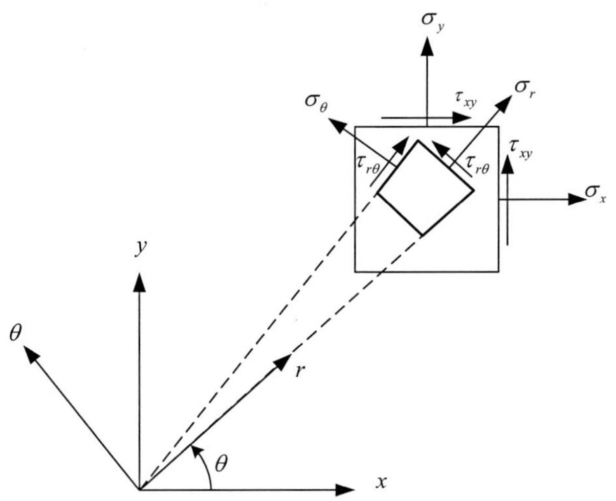

Fig. 1.10 Sketch illustrating the relation between the Cartesian and cylindrical components of stresses.

1.3.4 Equilibrium

For a 2-dimensional case (Fig. 1.11), the sum of all the surface-force components acting parallel to the x-axis on each outer plane of an infinitesimal material element of rectangular shape, and the corresponding body force component is

$$\int_V \left(\frac{\partial \sigma_x}{\partial x} + \frac{\partial \tau_{xy}}{\partial y} \right) dx dy, \quad \text{and} \quad \int_V F_x dx dy$$

respectively. The sum of these two must be the same as the mass times the acceleration of the material element, following the Newton's second law. Thus, we arrive at the following for the equation representing the force balance along the x-direction:

$$\frac{\partial \sigma_x}{\partial x} + \frac{\partial \tau_{xy}}{\partial y} + F_x - \rho \frac{d^2 u}{dt^2} = 0$$

where ρ is the density of the material. In a similar fashion, we can derive the y-component equation:

$$\frac{\partial \tau_{xy}}{\partial x} + \frac{\partial \sigma_y}{\partial y} + F_y - \rho \frac{d^2 v}{dt^2} = 0$$

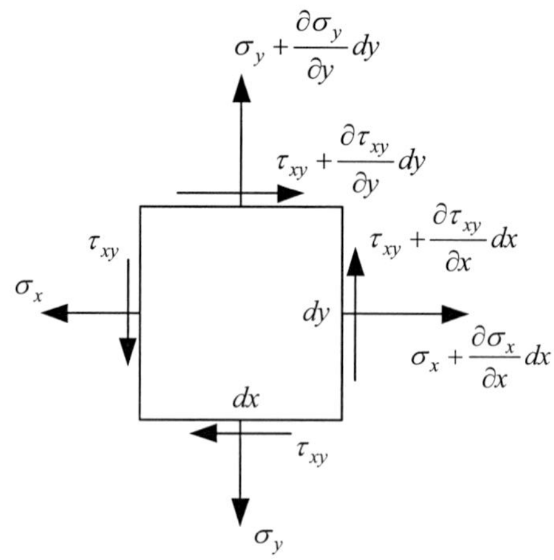

Fig. 1.11 Normal (σ) and tangential (τ) stresses acting on the whole surfaces of an infinitesimal 2-D material element.

For a 3-dimensional case, the equation takes the form

$$\frac{\partial \sigma_{ij}}{\partial x_j} + F_i = \rho \frac{d^2 u_i}{dt^2} \quad (i=1,\ 2,\ 3) \tag{1.3}$$

In a static state, the inertia term on the right-hand side vanishes and we get the equilibrium equations

$$\frac{\partial \sigma_{ij}}{\partial x_j} + F_i = 0 \quad (i=1,\ 2,\ 3) \tag{1.4}$$

We see that the number of equations is three whereas the number of unknowns is six, i.e., σ_{11}, σ_{22}, σ_{33}, σ_{12}, σ_{13}, σ_{23}; in fact, applying the principle of the conservation of angular momentum leads to the symmetric property of the stress tensor, i.e., $\sigma_{ij} = \sigma_{ji}$. Thus we need supplementary equations to close the problem.

In cylindrical coordinates, the equilibrium equations take the following form.

$$\frac{\partial \sigma_r}{\partial r} + \frac{1}{r}\frac{\partial \tau_{r\theta}}{\partial \theta} + \frac{\partial \tau_{rz}}{\partial z} + \frac{1}{r}(\sigma_r - \sigma_\theta) + F_r = 0$$

$$\frac{\partial \tau_{r\theta}}{\partial r} + \frac{1}{r}\frac{\partial \sigma_\theta}{\partial \theta} + \frac{\partial \tau_{\theta z}}{\partial z} + \frac{2}{r}\tau_{r\theta} + F_\theta = 0$$

$$\frac{\partial \tau_{rz}}{\partial r} + \frac{1}{r}\frac{\partial \tau_{\theta z}}{\partial \theta} + \frac{\partial \sigma_z}{\partial z} + \frac{1}{r}\tau_{rz} + F_z = 0$$

In spherical coordinates, the equilibrium equations take the following form.

$$\frac{\partial \sigma_R}{\partial R} + \frac{1}{R}\frac{\partial \tau_{R\phi}}{\partial \phi} + \frac{1}{R\sin\phi}\frac{\partial \tau_{R\theta}}{\partial \theta} + \frac{1}{R}(2\sigma_R - \sigma_\phi - \sigma_\theta + \tau_{R\phi}\cot\phi) + F_R = 0$$

$$\frac{\partial \tau_{R\phi}}{\partial R} + \frac{1}{R}\frac{\partial \sigma_\phi}{\partial \phi} + \frac{1}{R\sin\phi}\frac{\partial \tau_{\phi\theta}}{\partial \theta} + \frac{1}{R}[(\sigma_\phi - \sigma_\theta)\cot\phi + 3\tau_{R\phi}] + F_\phi = 0$$

$$\frac{\partial \tau_{R\theta}}{\partial R}+\frac{1}{R}\frac{\partial \tau_{\phi\theta}}{\partial \phi}+\frac{1}{R\sin\phi}\frac{\partial \sigma_\theta}{\partial \theta}+\frac{1}{R}(2\tau_{\phi\theta}\cot\phi+3\tau_{R\theta}]+F_\theta=0$$

1.4 Constituent Relations

1.4.1 Hooke's law for linear elastic materials

In order to attain the supplementary equations for the problem closure, we need relation between the stress tensor and the strain tensor. The most general form of this under the assumption of linear relation would be

$$\sigma_{ij}=C_{ijkl}e_{kl}$$

where C_{ijkl} is the forth-order elasticity tensor. Note that the number of equations is six. For isotropic materials, it has been shown that this simplifies to the generalized Hooke's law

$$\sigma_{ij}=\lambda e_{kk}\delta_{ij}+2\mu e_{ij} \tag{1.5}$$

where the *Lame's constant* λ and the *shear modulus* (or *modulus of rigidity*) μ are given as

$$\mu=\frac{E}{2(1+\nu)}$$

$$\lambda=\frac{\nu E}{(1+\nu)(1-2\nu)}$$

Here, E is the *modulus of elasticity* (or *Young's modulus*) and ν is the *Poisson's ratio.* In Cartesian coordinates, Eq. (1.5) becomes

$$\sigma_x=\lambda(e_x+e_y+e_z)+2\mu e_x$$

$$\sigma_y=\lambda(e_x+e_y+e_z)+2\mu e_y$$

$$\sigma_z=\lambda(e_x+e_y+e_z)+2\mu e_z$$

$$\tau_{xy}=2\mu e_{xy}$$

$$\tau_{yz}=2\mu e_{yz}$$

$$\tau_{zx}=2\mu e_{zx}$$

which may be called the 'strain-stress' relation. These forms are invariant of the coordinates selected. For instance, with the cylindrical coordinates, we can write

$$\sigma_r=\lambda(e_r+e_\theta+e_z)+2\mu e_r$$

etc.

Inverse of the above relation gives the 'stress-strain' relation reading

$$e_{ij} = \frac{1+\nu}{E}\sigma_{ij} - \frac{\nu}{E}\sigma_{kk}\delta_{ij} \tag{1.6}$$

Homework

Using Eqs. (1.1a) and (1.5), derive the equations relating the stress tensor σ_{ij} and the displacement vector u_i.

1.4.2 Meaning of modulus and Poisson ratio

For a simple tension force F acting along the x-direction on a bar with uniform area A (Fig. 1.12), we have $\sigma_x = \sigma_0$, $\sigma_y = \tau_{xy} = 0$, where $\sigma_0 = F/A$ is the applied stress. Following the stress-strain relations, we get

$$e_x = \frac{1+\nu}{E}\sigma_x - \frac{\nu}{E}\sigma_x = \frac{1}{E}\sigma_0$$

$$e_y = -\frac{\nu}{E}\sigma_0,\ e_z = -\frac{\nu}{E}\sigma_0$$

$$e_{xy} = 0$$

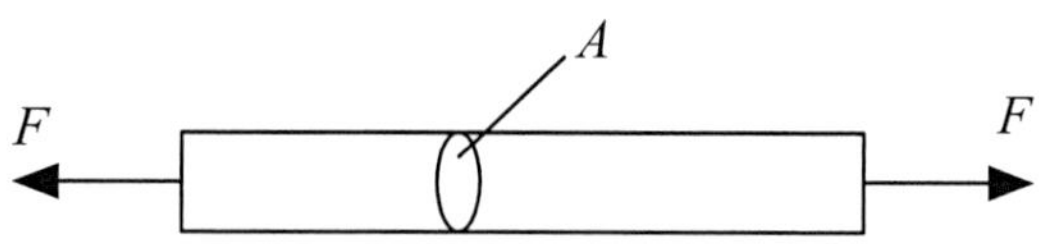

Fig. 1.12 A bar with a uniform cross-section under the tensile force.

The first result presents the 1-dimensional Hooke's law, $\sigma_x = Ee_x$, implying that E should be understood as the slope of the stress-strain curve. On the other hand, the second and third results indicate that the Poisson's ratio corresponds to the ratio of the (shrunk) principle strain normal to the x-axis and the (elongated) strain along the x-axis. For an incompressible material, we must require $e_x + e_y + e_z = 0$, which is achieved with $\nu = 0.5$.

For a pure shear load shown in Fig. 1.13, we have

$$\sigma_x = \sigma_y = \sigma_z = 0,\ \tau_{xy} = F/A = \tau_0,\ \tau_{yz} = \tau_{zx} = 0$$

Then, the stress-strain relations give

$$e_x = e_y = e_z = 0$$

$$e_{xy} = \frac{1+\nu}{E}\tau_{xy} = \frac{1+\nu}{E}\tau_0 = 2\mu\tau_0$$

$$e_{yz} = e_{zx} = 0$$

Thus, 2μ corresponds to the slope of the shear stress-strain curve, that is why μ is called *shear modulus*.

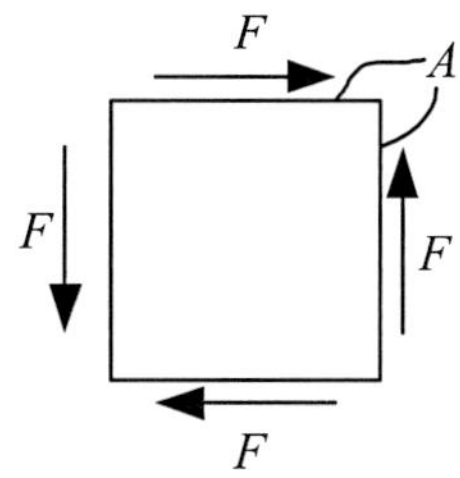

Fig. 1.13 The cross-section of a bar receiving pure shear forces.

For a hydrostatic compression or tension with the pressure p, the radius of the Mohr circle is zero and $\sigma_x = \sigma_y = \sigma_z = -p$, $\tau_{xy} = \tau_{yz} = \tau_{zx} = 0$. Then, the stress-strain relations give

$$e_x = e_y = e_z = \frac{-(1+\nu)}{E}p + \frac{3\nu}{E}p = \frac{-(1-2\nu)}{E}p$$

each of which corresponds to the linear strain of compression. The volumetric strain (amount of volume change per unit volume under the pressure) is e_{kk}, and thus we have $e_{kk} = -[3(1-2\nu)/E]p$, or

$$p = -ke_{kk}$$

where

$$k = \frac{E}{3(1-2\nu)}$$

is called the *bulk modulus*. This result indicates that an incompressible solid, $\nu = 0.5$, yields the bulk modulus infinitely large, which is reasonable.

There are many situations where the stress and the strain are affected by the variation of temperature. Take T_0 as the reference temperature and α as the thermal linear expansion coefficient. Then, contribution of the temperature change from T_0 to T to the shear strain is zero and that to the normal strain, $\alpha(T-T_0)$, will be the same for all the strain components. Thus, we have

$$e_{ij} = \frac{1+\nu}{E}\sigma_{ij} - \frac{\nu}{E}\sigma_{kk}\delta_{ij} + \alpha(T-T_0)\delta_{ij} \qquad (1.7)$$

as the modified stress-strain relation. Inversion of this results in the strain-stress relation reading

$$\sigma_{ij} = \lambda e_{kk}\delta_{ij} + 2\mu e_{ij} - (3\lambda + 2\mu)\alpha(T-T_0)\delta_{ij} \qquad (1.8)$$

This implies that increase of the temperature results in the decrease of the stress for the same strain.

1.5 Formulation and Boundary Conditions

1.5.1 Collection of equations defined or derived so far

For linear isotropic elasticity, the index form of the relevant equations in the Cartesian coordinates are:

Force balance equations:

$$\frac{\partial \sigma_{ij}}{\partial x_j}+F_i=0 \qquad \text{(3 equations)} \tag{1.9}$$

Displacement-strain relation:

$$e_{ij}=\frac{1}{2}\left(\frac{\partial u_i}{\partial x_j}+\frac{\partial u_j}{\partial x_i}\right) \qquad \text{(6 equations)} \tag{1.10}$$

Strain-stress and stress-strain (constitute) equations:

$$\sigma_{ij}=\lambda e_{kk}\delta_{ij}+2\mu e_{ij} \qquad \text{(6 equations)} \tag{1.11a}$$

$$e_{ij}=\frac{1+\nu}{E}\sigma_{ij}-\frac{\nu}{E}\sigma_{kk}\delta_{ij} \qquad \text{(6 equations)} \tag{1.11b}$$

So, we have 15 unknowns (3 of u_i, 6 of σ_{ij} and 6 of e_{ij}) with 15 equations (3 of force balance in Eq. (1.9), 6 of displacement-strain relation in Eq. (1.10), and 6 of either strain-stress (1.11a) or stress-strain relation (1.11b)).

○ In cylindrical coordinates:

Force balance equations:

$$\frac{\partial \sigma_r}{\partial r}+\frac{1}{r}\frac{\partial \tau_{r\theta}}{\partial \theta}+\frac{\partial \tau_{rz}}{\partial z}+\frac{1}{r}(\sigma_r-\sigma_\theta)+F_r=0$$

$$\frac{\partial \tau_{r\theta}}{\partial r}+\frac{1}{r}\frac{\partial \sigma_\theta}{\partial \theta}+\frac{\partial \tau_{\theta z}}{\partial z}+\frac{2}{r}\tau_{r\theta}+F_\theta=0$$

$$\frac{\partial \tau_{rz}}{\partial r}+\frac{1}{r}\frac{\partial \tau_{\theta z}}{\partial \theta}+\frac{\partial \sigma_z}{\partial z}+\frac{1}{r}\tau_{rz}+F_z=0$$

Displacement-strain relation:

$$e_r=\frac{\partial u_r}{\partial r},\quad e_\theta=\frac{1}{r}\left(u_r+\frac{\partial u_\theta}{\partial \theta}\right),\quad e_z=\frac{\partial u_z}{\partial z}$$

$$e_{r\theta}=\frac{1}{2}\left(\frac{1}{r}\frac{\partial u_r}{\partial \theta}+\frac{\partial u_\theta}{\partial r}-\frac{u_\theta}{r}\right)$$

$$e_{z\theta}=\frac{1}{2}\left(\frac{1}{r}\frac{\partial u_z}{\partial \theta}+\frac{\partial u_\theta}{\partial z}\right)$$

$$e_{zr}=\frac{1}{2}\left(\frac{\partial u_r}{\partial z}+\frac{\partial u_z}{\partial r}\right)$$

The strain-stress and the stress-strain relations take the same form as the one for the Cartesian coordinate system.

○ In spherical coordinates:

Force balance equations:

$$\frac{\partial \sigma_R}{\partial R}+\frac{1}{R}\frac{\partial \tau_{R\phi}}{\partial \phi}+\frac{1}{R\sin\phi}\frac{\partial \tau_{R\theta}}{\partial \theta}+\frac{1}{R}(2\sigma_R-\sigma_\phi-\sigma_\theta+\tau_{R\phi}\cot\phi)+F_R=0$$

$$\frac{\partial \tau_{R\phi}}{\partial R}+\frac{1}{R}\frac{\partial \sigma_\phi}{\partial \phi}+\frac{1}{R\sin\phi}\frac{\partial \tau_{\phi\theta}}{\partial \theta}+\frac{1}{R}[(\sigma_\phi-\sigma_\theta)\cot\phi+3\tau_{R\phi}]+F_\phi=0$$

$$\frac{\partial \tau_{R\theta}}{\partial R}+\frac{1}{R}\frac{\partial \tau_{\phi\theta}}{\partial \phi}+\frac{1}{R\sin\phi}\frac{\partial \sigma_\theta}{\partial \theta}+\frac{1}{R}(2\tau_{\phi\theta}\cot\phi+3\tau_{R\theta})+F_\theta=0$$

Displacement-strain relation:

$$e_R=\frac{\partial u_R}{\partial R},\ e_\phi=\frac{1}{R}\left(u_R+\frac{\partial u_\phi}{\partial \phi}\right),\ e_\theta=\frac{1}{R\sin\phi}\left(\frac{\partial u_\theta}{\partial \theta}+u_R\sin\phi+u_\phi\cos\phi\right)$$

$$e_{R\phi}=\frac{1}{2}\left(\frac{1}{R}\frac{\partial u_R}{\partial \phi}+\frac{\partial u_\phi}{\partial R}-\frac{u_\phi}{R}\right)$$

$$e_{\phi\theta}=\frac{1}{2R}\left(\frac{1}{\sin\phi}\frac{\partial u_\phi}{\partial \theta}+\frac{\partial u_\theta}{\partial \phi}-u_\theta\cot\phi\right)$$

$$e_{r\theta}=\frac{1}{2}\left(\frac{1}{R\sin\phi}\frac{\partial u_R}{\partial \theta}+\frac{\partial u_\theta}{\partial R}-\frac{u_\theta}{R}\right)$$

The strain-stress and the stress-strain relations take the same form as the ones for the Cartesian coordinate system.

1.5.2 Boundary conditions and choice of equations

In solving an elastic problem, appropriate equations to be chosen as the governing equations depend on the type of boundary conditions. When only surface tractions are given (including the case where boundaries are set free leading to zero stress), the stress equations must be employed. When the boundaries are given with strains or displacement (including the case with fixed boundaries leading to zero displacement there), the equations written in terms of the strain or those of the displacement vector are more appropriate.

1.5.3 Equations for the displacement

Substituting the displacement-strain relation into the strain-stress relation yields

$$\sigma_{ij}=\lambda\frac{\partial u_k}{\partial x_k}\delta_{ij}+\mu\left(\frac{\partial u_i}{\partial x_j}+\frac{\partial u_j}{\partial x_i}\right) \qquad (1.12)$$

Substituting this into the force balance (equilibrium) equation yields

$$\mu \frac{\partial^2 u_i}{\partial x_j \partial x_j} + (\lambda+\mu)\frac{\partial}{\partial x_i}\left(\frac{\partial u_k}{\partial x_k}\right) + F_i = 0 \tag{1.13}$$

In a vector form, this becomes

$$\mu \nabla^2 \boldsymbol{u} + (\lambda+\mu)\nabla(\nabla \cdot \boldsymbol{u}) + \boldsymbol{F} = 0 \tag{1.14}$$

which is called *Navier's equation* or *Lame's equation*. For an incompressible solid, $\nu = 0.5$, we have $\nabla \cdot \boldsymbol{u} = 0$ and Eq. (1.14) simplifies to

$$\mu \nabla^2 \boldsymbol{u} + \boldsymbol{F} = 0$$

1.6 Formulation of Two-Dimensional Problems

1.6.1 Plain strain problem

Considered in this category of 2-dimensional problems is an infinitely long cylinder with uniform cross-section and zero strains at both ends of the cylinder. Then we can assume $e_z = e_{zy} = e_{zx} = 0$ and all the variables depend only on the in-plane coordinates x and y. Then the displacement-strain relations simplify to

$$e_x = \frac{\partial u}{\partial x},\ e_y = \frac{\partial v}{\partial y},\ e_{xy} = \frac{1}{2}\left(\frac{\partial u}{\partial y} + \frac{\partial v}{\partial x}\right)$$

and the strain-stress relation to

$$\sigma_x = \lambda(e_x + e_y) + 2\mu e_x$$

$$\sigma_y = \lambda(e_x + e_y) + 2\mu e_y$$

$$\sigma_z = \lambda(e_x + e_y)$$

$$\tau_{xy} = 2\mu e_{xy}$$

$$\tau_{xz} = \tau_{yz} = 0$$

Note that the out-plane stress component σ_z becomes non-zero when $e_x + e_y$ is non-zero.

Now, the equations to be solved for the stress components are Eq. (1.9) or

$$\frac{\partial \sigma_x}{\partial x} + \frac{\partial \tau_{xy}}{\partial y} + F_x = 0 \tag{1.15a}$$

$$\frac{\partial \tau_{xy}}{\partial x} + \frac{\partial \sigma_y}{\partial y} + F_y = 0 \tag{1.15b}$$

We need one more equation in order to close the problem. For this, we can use Saint-Venant Eq. (1.2a):

$$\frac{\partial^2 e_x}{\partial y^2}+\frac{\partial^2 e_y}{\partial x^2}=2\frac{\partial^2 e_{xy}}{\partial x \partial y} \tag{1.16}$$

To replace the strains with the stresses, we use Eq. (1.11b):

$$e_x=\frac{1+\nu}{E}\left[(1-\nu)\sigma_x-\nu\sigma_y\right]$$

$$e_y=\frac{1+\nu}{E}\left[(1-\nu)\sigma_y-\nu\sigma_x\right]$$

$$e_{xy}=\frac{1+\nu}{E}\tau_{xy}$$

Substituting these into (1.16) yields

$$\nabla^2(\sigma_x+\sigma_y)=-\frac{1}{1-\nu}\nabla\cdot\boldsymbol{F} \tag{1.17}$$

where now ∇ is the two-dimensional gradient operator.

1.6.2 Plain stress problem

Here, the stress components are given (usually zero) at both ends of the cylinder, and the 2-dimensional assumption still holds. Then, we can assume $\sigma_z=\tau_{zy}=\tau_{zx}=0$. From Eq. (1.11b), we write

$$e_x=\frac{1}{E}(\sigma_x-\nu\sigma_y) \tag{1.18a}$$

$$e_y=\frac{1}{E}(\sigma_y-\nu\sigma_x) \tag{1.18b}$$

$$e_z=-\frac{\nu}{E}(\sigma_x+\sigma_y)=-\frac{\nu}{1-\nu}(e_x+e_y) \tag{1.18c}$$

$$e_{xy}=\frac{1+\nu}{E}\tau_{xy} \tag{1.18d}$$

Here, the second equality of Eq. (1.18c) comes from setting $\sigma_z=0$ in Eq. (1.11a) with $i=j=3$. For the governing equations with stresses as variables, we can still use Eqs. (1.15a) and (1.15b). The supplementary equation can be obtained from the Saint-Venant equation (1.16). Substituting Eqs. (1.18a), (1.18b) and (1.18d) into (1.16), we arrive at

$$\nabla^2(\sigma_x+\sigma_y)=-(1+\nu)\nabla\cdot\boldsymbol{F} \tag{1.19}$$

where the term $\partial^2\tau_{xy}/\partial x\partial y$ has been eliminated from the result of summing the derivative of (1.15a) with respect to x and the derivative of (1.15b) with respect to y.

To derive the equations of displacement, we write Eq. (1.13) with, e.g., $i=1$:

$$\mu\left(\frac{\partial^2 u}{\partial x^2}+\frac{\partial^2 u}{\partial y^2}\right)+(\lambda+\mu)\frac{\partial}{\partial x}\left(\frac{\partial u}{\partial x}+\frac{\partial v}{\partial y}+\frac{\partial w}{\partial z}\right)+F_x=0 \tag{1.20}$$

Note that although the 2-dimensional assumption allows to set $\partial^2 u/\partial z^2=0$, the term $\partial w/\partial z$ cannot be set zero. We write Eq. (1.12) with $i=j=3$ and set $\sigma_z=0$ to get

$$\lambda\left(\frac{\partial u}{\partial x}+\frac{\partial v}{\partial y}+\frac{\partial w}{\partial z}\right)+2\mu\frac{\partial w}{\partial z}=0$$

or

$$\frac{\partial w}{\partial z}=-\frac{\lambda}{\lambda+2\mu}\left(\frac{\partial u}{\partial x}+\frac{\partial v}{\partial y}\right)$$

Substituting this into Eq. (1.20) we arrive at

$$\mu\left(\frac{\partial^2 u}{\partial x^2}+\frac{\partial^2 u}{\partial y^2}\right)+\frac{E}{2(1-\nu^2)}\frac{\partial}{\partial x}\left(\frac{\partial u}{\partial x}+\frac{\partial v}{\partial y}\right)+F_x=0$$

In a similar fashion, we can derive the corresponding equation for v, i.e., $i=2$. The final vector form of the equation governing the displacement is

$$\mu\nabla^2\boldsymbol{u}+\frac{E}{2(1-\nu^2)}\nabla(\nabla\cdot\boldsymbol{u})+\boldsymbol{F}=0 \tag{1.21}$$

It can be seen that the formulation for the plane strain and that for the plane stress take the same form except constants; compare Eqs. (1.17) and (1.19), and compare Eqs. (1.14) and (1.21).

1.6.3 Airy stress function

For the plane strain and plane stress problems, we can apply the method of Airy stress function. When the body force is conservative, a potential V can be found such that $\boldsymbol{F}=-\nabla V$. Further, we introduce the Airy function $\phi(x,y)$ defined through the following equations:

$$\sigma_x=\frac{\partial^2\phi}{\partial y^2}+V \tag{1.22a}$$

$$\sigma_y=\frac{\partial^2\phi}{\partial x^2}+V \tag{1.22b}$$

$$\tau_{xy}=-\frac{\partial^2\phi}{\partial x\partial y} \tag{1.22c}$$

Then, the equilibrium equations, (1.15a) and (1.15b), are automatically satisfied. The remaining equations to be satisfied are the compatability equations, i.e., Eq. (1.17) for the plane strain and Eq. (1.19) for the plane stress case. Substituting the above equations for the stress components into (1.17) and (1.19) produces

$$\nabla^2(\nabla^2\phi)=-\frac{1-2\nu}{1-\nu}\nabla^2 V \quad \text{(plane strain)} \tag{1.23a}$$

$$\nabla^2(\nabla^2\phi)=-(1-\nu)\nabla^2 V \quad \text{(plane stress)} \tag{1.23b}$$

If the external load is zero, $V=\text{constant}$, we get the bi-harmonic equation

$$\nabla^2(\nabla^2\phi)=0 \tag{1.24}$$

as the governing equations for both the plane strain and plane stress problems.

Homework

For the cylindrical coordinate system,

(1) simplify the 3-dimensional equilibrium equations to 2-dimensional form and

(2) derive the compatibility equations for both the plane strain and plane stress problems.

(3) In each case, starting from Eqs. (1.17) and (1.19) respectively, show that the following setting

$$\sigma_r=\frac{1}{r}\frac{\partial\phi}{\partial r}+\frac{1}{r^2}\frac{\partial^2\phi}{\partial\theta^2},\quad \sigma_\theta=\frac{\partial^2\phi}{\partial r^2},\quad \tau_{r\theta}=-\frac{\partial}{\partial r}\left(\frac{1}{r}\frac{\partial\phi}{\partial\theta}\right)$$

automatically satisfies the force equilibrium equations for the case without the body force.

(4) Substitute these into the compatibility equations to get the final equations to be solved for the Airy stress function ϕ in the cylindrical coordinates.

1.7 Solutions of Two-Dimensional Problems

1.7.1 Using polynomials with Cartesian coordinates

Even for a 2-dimensional problem, the governing equations take different forms depending on the end conditions; i.e., the ends may be set either strain-free (plane strain problem) or stress-free (plane stress problem); specifically, the constants multiplying the external force terms are different. In this chapter, however, we restrict ourselves to the plane stress problems.

Regardless of the end conditions, in the absence of the body force, the equation to be solved is the bi-harmonic equation (1.24). For the case when the Cartesian coordinates are appropriate, one may expand the Airy function in a power series of x and y:

$$\phi(x,y)=\sum_{m=0}^{\infty}\sum_{n=0}^{\infty}A_{mn}x^m y^n \tag{1.25}$$

Substituting this into Eq. (1.24), we get the recursive formula for the coefficients:

$$(m+2)(m+1)m(m-1)A_{m+2,n-2}+2m(m-1)n(n-1)A_{mn} \\ +(n+2)(n+1)n(n-1)A_{m-2,n+2}=0 \tag{1.26}$$

We can see that terms with $m+n \leq 3$, say $A_{11}xy$ and $A_{03}y^3$ etc, automatically satisfy the bi-harmonic equation without restriction to the coefficients. Depending on the problem, we are allowed to use the truncated series.

Consider a uniform tension of a beam caused by an externally applied tensional stress T as shown in Fig. 1.14.

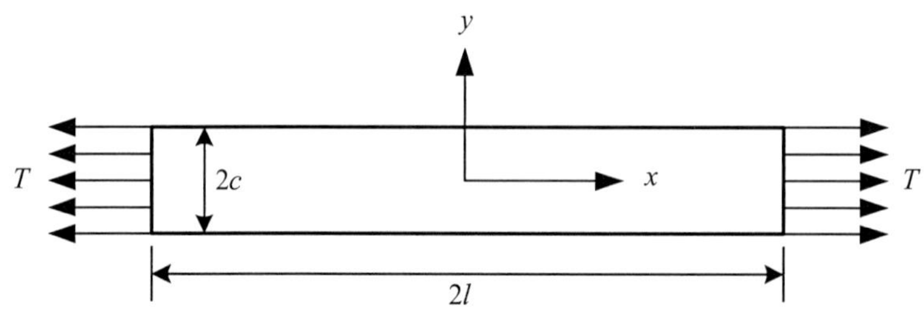

Fig. 1.14 A beam receiving pure tensile force T.

The boundary conditions are

$$\sigma_x(\pm l, y) = T, \quad \sigma_y(x, \pm c) = 0, \quad \tau_{xy}(\pm l, y) = \tau_{xy}(x, \pm c) = 0$$

As shown in Eqs. (1.22a) and (1.22b), since the stresses are second derivatives of the Airy function, we try the function $\phi = A_{02}y^2$. Applying the boundary conditions, we get $A_{02} = T/2$, and $\phi = Ty^2/2$, from which we arrive at

$$\sigma_x = T, \quad \sigma_y = \tau_{xy} = 0$$

The displacement components u and v can be given from (1.1a), (1.18a) and (1.18b). First we get, from the definition of e_x and e_y, (1.18a) and (1.18b), the followings.

$$e_x = T/E, \quad e_y = -\nu T/E$$

We then integrate Eq. (1.1a) for e_x and e_y to obtain

$$u = \frac{T}{E}x + f(y), \quad v = -\nu\frac{T}{E}y + g(x)$$

The two unknown functions $f(y)$ and $g(x)$ are to be determined from the formula for e_{xy}, i.e.,

$$e_{xy} = \frac{1}{2\mu}\tau_{xy} = \frac{1}{2}\left(\frac{\partial u}{\partial y} + \frac{\partial v}{\partial x}\right)$$

Substituting for the equations of u and v on the right-hand side and setting $\tau_{xy} = 0$ leads to

$$f'(y) + g'(x) = 0$$

Therefore, we must set $g'(x) = -f'(y) = \text{constant} \equiv \omega_0$. In fact, the constant ω_0 adjusts the amount of rotation experienced by the infinitesimal material element, which is free to choose; see Eq. (1.1b). Thus we have

$$f(y) = -\omega_0 y + u_0$$

$$g(x) = \omega_0 x + v_0$$

Here, u_0 and v_0 correspond to the components of the rectilinear displacement of the element. We can set ω_0, u_0 and v_0 in such way that the resultant displacement vector can fit the specified boundary conditions.

Homework

In the above problem, replace the tension at the end sections, $x = \pm l$, by an end moment, M (Fig. 1.15). Construct suitable boundary conditions and solve the problem by using the form, $\phi = A_{03} y^3$. In setting the integration constants for the displacement components, apply the restriction of no displacement and no rotation at $x = y = 0$.

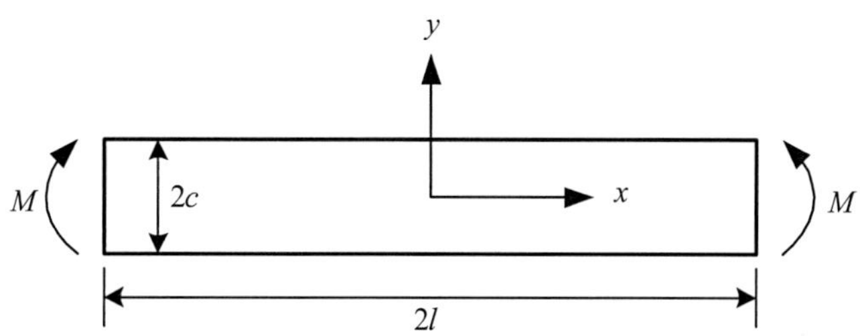

Fig. 1.15 A beam receiving pure moment M at both ends.

Next, we consider bending of a beam of uniform cross section by a uniform distribution of load w (Fig. 1.16).

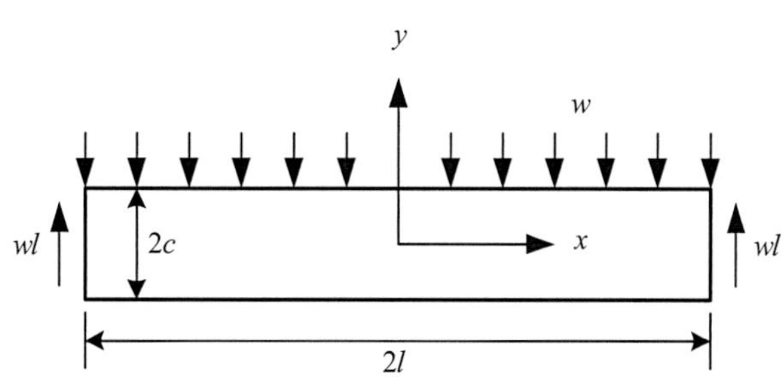

Fig. 1.16 A beam receiving uniformly distributed side force per length, w.

The boundary conditions are

$\sigma_y(x, c) = -w$ (uniform normal load on top) (1.27a)

$\sigma_y(x, -c) = 0$ (zero normal load on bottom) (1.27b)

$\tau_{xy}(x, \pm c) = 0$ (zero shear load on top and bottom) (1.27c)

$$\int_{-c}^{c} \sigma_x(\pm l, y)dy = 0 \text{ (zero normal force on both side walls)} \quad (1.27d)$$

$$\int_{-c}^{c} \sigma_x(\pm l, y)ydy = 0 \text{ (zero moment on both side walls)} \quad (1.27e)$$

$$\int_{-c}^{c} \tau_{xy}(\pm l, y)dy = \pm wl \text{ (force balance with applied load on side walls)} \quad (1.27f)$$

Note that the boundary conditions on the side walls are given in the integral form, not in a pointwise manner, because we intend to take simple polynomials as the solutions. Since the normal stress conditions on the top and bottom surfaces are free of the coordinate x, we must assume that σ_y should also be free of x, say only a function of y; $\sigma_y = f(y)$. Integrating this twice with respect to x, we get the Airy function ϕ. So, we expect that ϕ must show a second-order polynomial with respect to x; but the linear term should be excluded because of the symmetric property of the solutions with respect to the axis $x=0$. Since the boundary conditions on the side walls are given in the integral form, we are free to choose the functional dependence of the stresses on the coordinate y. Next, in order to derive the correct functional form of σ_y with respect to y, we think the boundary condition for σ_y on the top wall as

$$\sigma_y(x,c) = -w/2 - w/2$$

$$\sigma_y(x,-c) = -w/2 + w/2$$

The first parts on the right-hand side contribute to the uniform distribution of σ_y in y, whereas the second parts are related to the linear or cubic polynomials with respect to y of the solutions. So, we can assume $\sigma_y = C_1 + C_2 y + C_3 y^3$, where C_1, C_2 and C_3 are real constants. Integrating this twice with respect to x gives ϕ. Thus, we can take the following form for ϕ.

$$\phi = A_{20}x^2 + A_{21}x^2y + A_{03}y^3 + A_{23}x^2y^3 - \frac{A_{23}}{5}y^5$$

The final term has been added to satisfy the bi-harmonic equation or the recursive relation (1.26). The stress terms are given as follows.

$$\sigma_x = 6A_{03}y + 6A_{23}x^2y - 4A_{23}y^3$$

$$\sigma_y = 2A_{20} + 2A_{21}y + 2A_{23}y^3$$

$$\tau_{xy} = -2A_{21}x - 6A_{23}xy^2$$

Now, applying the first three boundary conditions, Eqs. (1.27a)-(1.27c), we get

$$-w = 2A_{20} + 2A_{21}c + 2A_{23}c$$

$$0 = 2A_{20} - 2A_{21}c - 2A_{23}c^3$$

$$0 = -2A_{21} - 6A_{23}c^2$$

These can be solved for three unknown coefficients:

$$A_{20} = -\frac{w}{4},\ A_{21} = -\frac{3w}{8c},\ A_{23} = \frac{w}{8c^3}$$

It can be seen that the boundary conditions (1.27d) and (1.27f) are satisfied. The remaining boundary condition (1.27e) is satisfied with

$$A_{03} = \frac{w}{8c^3}\left(\frac{2}{5}c^2 - l^2\right)$$

This completes the solution procedure. Now, we see that both σ_x and τ_{xy} are functions of x on the side walls, and hence the pointwise type of boundary conditions could not be employed there with the polynomial-type solutions.

Homework

Obtain the displacement vector field under the conditions of no rotation and no displacement at $x = y = 0$. Express the maximum vertical displacement in terms of w, l, E and $I = 2c^3/3$.

1.7.2 Using Fourier methods with Cartesian coordinates

In solving the bi-harmonic equation $\nabla^2(\nabla^2\phi) = 0$, or

$$\nabla^2\Phi = 0$$

$$\nabla^2\phi = \Phi$$

we can also try, as the general solution to the first equation, a Fourier form

$$\Phi = [C_1\cos\alpha x + C_2\sin\alpha x]\cosh(\alpha y) + [D_1\cos\alpha x + D_2\sin\alpha x]\sinh(\alpha y)$$

where α is a real constant which can be determined from the given domain's periodic nature; note that this form applies to the case where the solutions are periodic in x. Arbitrary constants C_1, C_2, D_1 and D_2 are to be determined from two sets of paired boundary conditions. Substituting these into the second equation, we get the general solution in the form similar to the above except that the coefficients are replaced by linear function of y as shown below.

$$\phi = [(C_{11} + C_{12}y)\cos\alpha x + (C_{21} + C_{22}y)\sin\alpha x]\cosh\alpha y$$
$$+ [(D_{11} + D_{12}y)\cos\alpha x + (D_{21} + D_{22}y)\sin\alpha x]\sinh\alpha y$$

where 8 coefficients are determined from 8 boundary conditions.

Suppose a sinusoidal distribution of load is applied on the top wall of a rectangular beam as shown in Fig. 1.17.

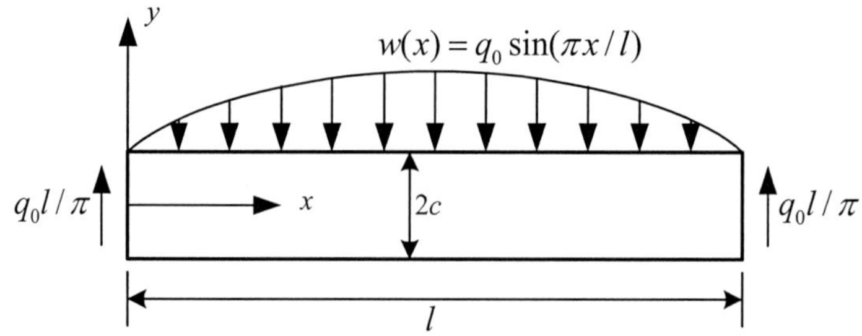

Fig. 1.17 A beam receiving non-uniformly distributed side force per length, $w(x)$.

The boundary conditions are:

$$\sigma_x(0,y) = 0, \quad \sigma_x(l,y) = 0 \tag{1.28a}$$

$$\tau_{xy}(x, \pm c) = 0 \tag{1.28b}$$

$$\sigma_y(x, -c) = 0, \tag{1.28c}$$

$$\sigma_y(x,c) = -q_0 \sin(\pi x / l) \tag{1.28d}$$

$$\int_{-c}^{c} \tau_{xy}(0,y)dy = -q_0 l/\pi, \quad \int_{-c}^{c} \tau_{xy}(l,y)dy = q_0 l/\pi \tag{1.28e}$$

Considering the functional form of the boundary condition (1.28d), we assume as the general solution the following form.

$$\phi = \sin\alpha x[(A + C\alpha y)\sinh\alpha y + (B + D\alpha y)\cosh\alpha y] \tag{1.29}$$

where $\alpha = \pi / l$. From this, we derive

$$\sigma_x = \alpha^2 \sin\alpha x[(A + 2D + C\alpha y)\sinh\alpha y + (B + 2C + D\alpha y)\cosh\alpha y]$$

$$\sigma_y = -\alpha^2 \sin\alpha x[(A + C\alpha y)\sinh\alpha y + (B + D\alpha y)\cosh\alpha y]$$

$$\tau_{xy} = -\alpha^2 \cos\alpha x[(A + D + C\alpha y)\cosh\alpha y + (B + C + D\alpha y)\sinh\alpha y]$$

We can see that the boundary condition (1.28a) is automatically satisfied. Applying the boundary condition (1.28b), we get

$$A\cosh\alpha c + D(\cosh\alpha c + \alpha c \sinh\alpha c) = 0$$

$$B\sinh\alpha c + C(\sinh\alpha c + \alpha c \cosh\alpha c) = 0$$

The boundary conditions (1.28c) and (1.28d) supply

$$(A - C\alpha c)\sinh\alpha c - (B - D\alpha c)\cosh\alpha c = 0$$

$$(A + C\alpha c)\sinh\alpha c + (B + D\alpha c)\cosh\alpha c = q_0 / \alpha^2$$

It can be shown that the boundary condition (1.28e) does not yield an independent equation; in fact, the third and forth equations derived above can be combined to yield the same equation as what would be derived from application of the boundary condition (1.28e). From the above 4 equations we can derive the unknown constants A, B, C, and D.

Homework

Obtain the displacement vector field with conditions of no rotation and no displacement at $x=l/2,\ y=0$. Derive the maximum vertical displacement.

For an arbitrary distribution of top load, we can construct a Fourier series as the solution's functional dependence on x. As an example, for the linearly distributed load (Fig. 1.18)

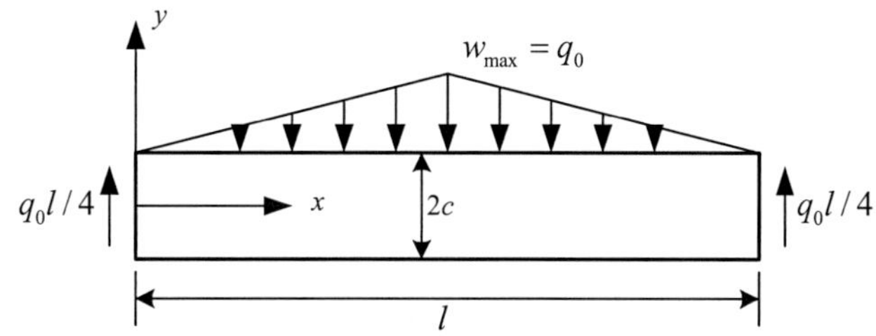

Fig. 1.18 A beam receiving linearly distributed side force per length, $w(x)$.

$$w(x)=\begin{cases}2q_0x/l & \text{for } 0\leqq x\leqq l/2\\ 2q_0(x/l-1) & \text{for } l/2<x\leqq l\end{cases}$$

we can transform this to the Fourier-expansion form

$$w(x)=\sum_{k=1}^{\infty}C_k\sin(\pi kx/l)$$

where the Fourier coefficients are given as

$$C_k=\frac{2}{l}\int_0^l w(x)\sin(\pi kx/l)dx$$

Then the remaining work for the solutions are straight-foward.

Homework

With suitable boundary conditions, obtain the solutions of stresses for the above problem.

1.7.3 Solutions with cylindrical coordinates

Now we consider 2-dimensional elasticity problems which can be effectively treated with polar coordinates. Again in the absence of the body force or for the case where the potential function V of the body force satisfies the Laplace equation, $\nabla^2 V=0$, the Airy stress function $\phi(r,\theta)$ must satisfy the bi-harmonic equation written now in the polar coordinates as follows.

$$(\nabla^2)^2\phi = \left(\frac{\partial^2}{\partial r^2} + \frac{1}{r}\frac{\partial}{\partial r} + \frac{1}{r^2}\frac{\partial^2}{\partial \theta^2}\right)^2 \phi = 0 \tag{1.30}$$

Assume a separable form $\phi = r^a \exp(b\theta)$, where a and b are arbitrary constants. Substituting this into the above equation leads to the characteristic equation

$$(a^2 + b^2)[(a-2)^2 + b^2] = 0$$

The roots are:

$$a = \pm ib, \quad a = 2 \pm ib$$

If the solution is periodic in θ, then b must be of the type, $b = in$, where n is an integer and a must take the form, $a = \pm n$ and $a = 2 \pm n$. Thus, the general solution can be written as

$$\phi = \sum_{n=0}^{\infty} \phi_n$$

where, for $n \geqq 1$,

$$\begin{aligned}\phi_n &= \left(a_{n1}r^n + a_{n2}r^{2+n} + a_{n3}r^{-n} + a_{n4}r^{2-n}\right)\cos n\theta \qquad (n = 1, 2, 3, \ldots.) \\ &+ \left(b_{n1}r^n + b_{n2}r^{2+n} + b_{n3}r^{-n} + b_{n4}r^{2-n}\right)\sin n\theta\end{aligned} \tag{1.31}$$

For $n = 0$ (axi-symmetric mode), we have $a = 0$ (double root) and $a = 2$ (double root). Therefore ϕ_0 takes the following form:

$$\phi_0 = a_0 + a_1 \ln r + a_2 r^2 + a_3 r^2 \ln r \tag{1.32}$$

The stress components in the polar coordinates are given as

$$\sigma_r = \frac{1}{r}\frac{\partial \phi}{\partial r} + \frac{1}{r^2}\frac{\partial^2 \phi}{\partial \theta^2} \tag{1.33a}$$

$$\sigma_\theta = \frac{\partial^2 \phi}{\partial r^2} \tag{1.33b}$$

$$\tau_{r\theta} = -\frac{\partial}{\partial r}\left(\frac{1}{r}\frac{\partial \phi}{\partial \theta}\right) \tag{1.33c}$$

Consider an axi-symmetric problem between two concentric cylinders as shown in Fig. 1.19. Pressure is applied on the inner and outer walls of the domain. We also assume the plain-strain condition.

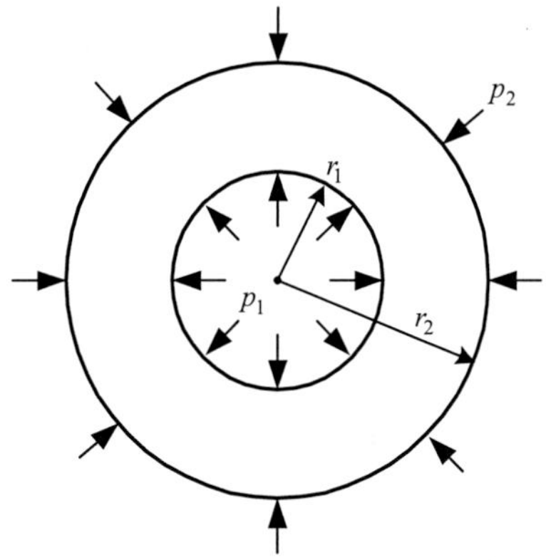

Fig. 1.19 A cylindrical shell receiving pressures from both inner and outer walls.

Since the solution should be axi-symmetric, the Airy function must take the form (1.32). Then the stress components are obtained from Eqs. (1.33a) - (1.33c):

$$\sigma_r = 2a_3 \ln r + \frac{a_1}{r^2} + a_3 + 2a_2$$

$$\sigma_\theta = 2a_3 \ln r - \frac{a_1}{r^2} + 3a_3 + 2a_2$$

$$\tau_{r\theta} = 0$$

For the plain-strain problem, the relation between the strain and stress components are

$$\sigma_r = \lambda(e_r + e_\theta) + 2\mu e_r$$

$$\sigma_\theta = \lambda(e_r + e_\theta) + 2\mu e_\theta$$

$$\tau_{r\theta} = 2\mu e_{r\theta}$$

From these, we get

$$e_r = \frac{1+\nu}{E}\left[(1-\nu)\sigma_r - \nu\sigma_\theta\right]$$

$$= \frac{1+\nu}{E}\left[2a_3(1-2\nu)\ln r + \frac{a_1}{r^2} + (1-4\nu)a_3 + 2a_2(1-2\nu)\right]$$

$$e_\theta = \frac{1+\nu}{E}\left[(1-\nu)\sigma_\theta - \nu\sigma_r\right]$$

$$= \frac{1+\nu}{E}\left[2a_3(1-2\nu)\ln r - \frac{a_1}{r^2} + (3-4\nu)a_3 + 2a_2(1-2\nu)\right]$$

$$e_{r\theta} = 0$$

Now, the displacement-strain relations in polar coordinates are

$$e_r = \frac{\partial u_r}{\partial r}, \quad e_\theta = \frac{1}{r}\left(u_r + \frac{\partial u_\theta}{\partial \theta}\right)$$

Integrating the first relation gives

$$u_r = \frac{1+\nu}{E}\left[2a_3(1-2\nu)r\ln r - \frac{a_1}{r} - a_3 r + 2a_2(1-2\nu)r\right]$$

Substituting this into the equation for e_θ and integrating the result with respect to θ yields

$$u_\theta = \frac{4(1-\nu^2)}{E}a_3 r\theta$$

Since the variable u_θ must be one-valued function, we require $a_3 = 0$. Then we are left with

$$\sigma_r = \frac{a_1}{r^2} + 2a_2, \quad \sigma_\theta = -\frac{a_1}{r^2} + 2a_2$$

The two constants can be obtained from application of boundary conditions at $r = r_1$ and $r = r_2$.

Homework

Derive the constants a_1 and a_2 from application of the pressure conditions at $r = r_1$ and $r = r_2$. Show that the formula for calculating the axial stress $\sigma_z = \nu(\sigma_r + \sigma_\theta)$ is $\sigma_z = 2\nu(r_1^2 p_1 - r_2^2 p_2)/(r_2^2 - r_1^2)$.

Consider an infinite medium with a stress-free cylindrical hole of radius a at the origin of the coordinate system under a uniform far-field tension loading with the normal stress T as shown in Fig. 1.20.

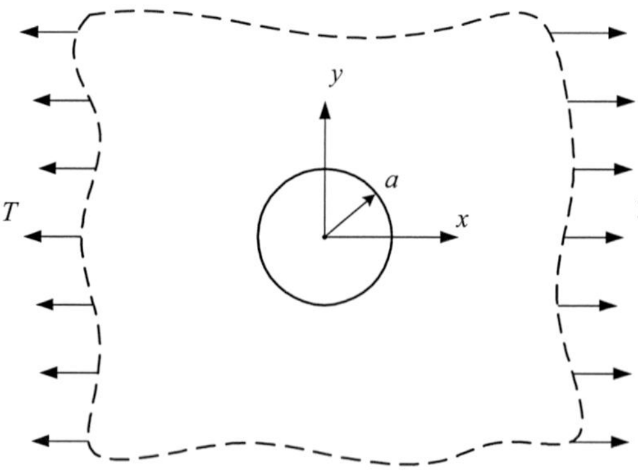

Fig. 1.20 An infinite medium having a cylindrical hole subjected to a horizontal tension force T.

The boundary conditions for the stress components $\sigma_r(r,\theta)$, $\sigma_\theta(r,\theta)$, and $\tau_{r\theta}(r,\theta)$ are

$$\sigma_r(a,\theta) = \tau_{r\theta}(a,\theta) = 0$$

$$\sigma_r(\infty,\theta) = \frac{T}{2}(1+\cos 2\theta)$$

$$\sigma_\theta(\infty,\theta) = \frac{T}{2}(1-\cos 2\theta)$$

$$\tau_{r\theta}(\infty,\theta) = -\frac{T}{2}\sin 2\theta$$

where the far-field conditions for σ_r and σ_θ are given from the transformation rule provided in Section 1.3.3. For instance, σ_r can be obtained from the formula for σ'_x and by setting $\sigma_x = T$, $\sigma_y = 0$, $\tau_{xy} = 0$, $\sigma'_x = \sigma_r$. Looking into the boundary conditions at infinity, we assume $\phi = \phi_0 + \phi_2$ as the general solution, where ϕ_2 contains only the term with $\cos 2\theta$ in Eq. (1.31):

$$\phi = a_0 + a_1 \ln r + a_2 r^2 + a_3 r^2 \ln r + \left(a_{21}r^2 + a_{22}r^4 + a_{23}r^{-2} + a_{24}\right)\cos 2\theta$$

To satisfy the stress conditions at infinity, we must require $a_3 = a_{22} = 0$. Stress components are given from (1.33a)-(1.33c):

$$\sigma_r = 2a_2 + \frac{a_1}{r^2} - \left(2a_{21} + \frac{6a_{23}}{r^4} + \frac{4a_{24}}{r^2}\right)\cos 2\theta$$

$$\sigma_\theta = 2a_2 - \frac{a_1}{r^2} + \left(2a_{21} + \frac{6a_{23}}{r^4}\right)\cos 2\theta$$

$$\tau_{r\theta} = \left(2a_{21} - \frac{6a_{23}}{r^4} - \frac{2a_{24}}{r^2}\right)\sin 2\theta$$

Applying the boundary conditions, we get

$$a_1 = -\frac{a^2 T}{2},\quad a_2 = \frac{T}{4},\quad a_{21} = -\frac{T}{4},\quad a_{23} = -\frac{a^4 T}{4},\quad a_{24} = \frac{a^2 T}{2}$$

The azimuthal stress component on the hole, $\sigma_\theta = T(1-2\cos 2\theta)$, has the maximum value $3T$ at the position $\theta = \pi/2$.

Homework

Repeat the solution process for an infinity medium with a stress-free hole at the origin of the coordinate system and now with the bi-axial loading, so that the stress components at infinity are: $\sigma_x = T$, $\sigma_y = -T$, $\tau_{xy} = 0$.

1.7.4 Solutions by using strain potential functions

Compared with the problems that are effectively solved by using Cartesian and cylindrical coordinates, the problems of spherical coordinates are more difficult even for axi-symmetric cases. Instead of using Airy stress functions, the axially symmetric spherical problems employ strain functions, such as Lame's strain potential and Love's strain potential functions. Substantial part of this Section comes from the text book of A.P.S. Selvadurai, "Partial Differential Equations in Mechanics 2" (Springer, 2000).

Lame's strain potential function φ is defined as

$$2\mu \boldsymbol{u} = \nabla\varphi \tag{1.34}$$

where $\boldsymbol{u}$ is the displacement vector. On the other hand, in the absence of the body force term, the displacement equation (1.14) reads

$$\nabla^2 \boldsymbol{u} + \frac{1}{(1-2\nu)}\nabla(\nabla\cdot\boldsymbol{u}) = 0$$

Substituting Eq. (1.34) into this, we get the requirement $\nabla(\nabla^2\varphi) = 0$ implying that

$$\nabla^2\varphi = C \tag{1.35}$$

where C is a constant. This constant corresponds to the homogeneous state of the stress, because divergence of $\boldsymbol{u}$ is proportional to the sum of normal stresses;

$$\nabla\cdot\boldsymbol{u} = \nabla^2\varphi/2\mu = e_{kk} = (1-2\nu)\sigma_{kk}/E$$

indicating that C is the same as $\sigma_{kk}(1-2\nu)/(1+\nu)$. So, the fundamental solution of φ is obtained from the Laplace equation $\nabla^2\varphi = 0$. In this sense, the Lame's strain potential function φ is called *harmonic*; this is contrary to the term *bi-harmonic* to be used shortly in denoting an arbitrary function f satisfying the bi-harmonic equation, $\nabla^4 f = 0$. Now, taking curl to Eq. (1.34), we see that $\nabla\times\boldsymbol{u} = 0$, because curl of gradient of any scalar function is always zero. This means that the Lame's strain potential function can only solve the problem that carries irrotational deformation, such as pure elongation (also pure contraction) along the principal axis. In order to tackle the rotational deformation case, we must use a more generalized method, one of which is the Love's strain potential method.

To understand the Love's strain potential, we should first understand the Galerkin vector potential. Galerkin proposed to use a vector potential $\boldsymbol{V}$ defined through

$$2\mu\boldsymbol{u} = 2(1-\nu)\nabla^2\boldsymbol{V} - \nabla(\nabla\cdot\boldsymbol{V}) \tag{1.36}$$

as a derived variable for use in solving the elasticity problems. Substituting this into (1.14), we end up with

$$\nabla^4\boldsymbol{V} = 0 \tag{1.37}$$

The reason why we adopt a new function like $\boldsymbol{V}$ instead of $\boldsymbol{u}$ in deriving the final form for the governing equations in solving practical problems can be seen from this simplified result; the reason is because the resultant equation takes a more simplified form and/or a well known form, so that well-developed mathematical tools can be effectively employed. In Cartesian coordinate system, $\boldsymbol{V}$ is expressed as

$\boldsymbol{V} = V_x\boldsymbol{i} + V_y\boldsymbol{j} + V_z\boldsymbol{k}$, and thus each component is bi-harmonic. However for the curved coordinates, each component is not bi-harmonic, because spatial derivatives of the unit vectors themselves do not vanish. For instance, with cylindrical coordinates, $\partial(\boldsymbol{i}_r)/\partial\theta = \boldsymbol{i}_\theta$, etc. In fact, it can be shown that for the cylindrical coordinate system with $\boldsymbol{V} = V_r\boldsymbol{i}_r + V_\theta\boldsymbol{i}_\theta + V_z\boldsymbol{i}_z$, the terms $\nabla^4 V_r$ and $\nabla^4 V_\theta$ are non-zero and turn out to take complicated forms (see Selvadurai, 2000). However, the term $\nabla^4 V_z$ is zero, because the unit vector $\boldsymbol{i}_z$ is invariant of the coordinates. Since our concern here is axi-symmetric problems which can be effectively treated with cylindrical or spherical coordinates, we present formulation in the cylindrical coordinates first and then the one in the spherical coordinates.

Love's strain potential Φ is defined as the z-component of the Galerkin's vector potential, that is, the magnitude of $\boldsymbol{V} = \Phi\boldsymbol{i}_z$ written in the cylindrical coordinate system. Then Φ is bi-harmonic in the sense that it satisfies the bi-harmonic equation

$$\nabla^4\Phi = 0 \tag{1.38}$$

Let us derive the relation between the Love strain potential Φ and the displacement vector in cylindrical coordinates. From Eq. (1.36), with $\boldsymbol{V} = \Phi\boldsymbol{i}_z$, we get each component of the displacement vector, $\boldsymbol{u} = u_r\boldsymbol{i}_r + u_\theta\boldsymbol{i}_\theta + u_z\boldsymbol{i}_z$, as follows.

$$2\mu u_r = -\frac{\partial^2\Phi}{\partial r\partial z}$$

$$2\mu u_\theta = -\frac{1}{r}\frac{\partial^2\Phi}{\partial\theta\partial z}$$

$$2\mu u_z = 2(1-\nu)\nabla^2\Phi - \frac{\partial^2\Phi}{\partial z^2}$$

where $\text{grad} = (\partial/\partial r,\ \partial/r\partial\theta,\ \partial/\partial z)$ and $\nabla\cdot\boldsymbol{F} = \partial(rF_r)/\partial r + \partial F_\theta/r\partial\theta + \partial F_z/\partial z$ for an arbitrary vector $\boldsymbol{F}$ are used. Substitute these into the displacement-strain relations and then the strains into the strain-stress relations presented in Section 1.5.1 in terms of cylindrical coordinates, we get

$$\sigma_r = \frac{\partial}{\partial z}\left[\nu\nabla^2\Phi - \frac{\partial^2\Phi}{\partial r^2}\right]$$

$$\sigma_\theta = \frac{\partial}{\partial z}\left[\nu\nabla^2\Phi - \frac{1}{r}\frac{\partial\Phi}{\partial r} - \frac{1}{r^2}\frac{\partial^2\Phi}{\partial\theta^2}\right]$$

$$\sigma_z = \frac{\partial}{\partial z}\left[(2-\nu)\nabla^2\Phi - \frac{\partial^2\Phi}{\partial z^2}\right]$$

$$\tau_{r\theta} = -\frac{\partial^3}{\partial r\,\partial\theta\,\partial z}\left[\frac{\Phi}{r}\right]$$

$$\tau_{\theta z} = \frac{1}{r}\frac{\partial}{\partial \theta}\left[(1-\nu)\nabla^2\Phi - \frac{\partial^2 \Phi}{\partial z^2}\right]$$

$$\tau_{rz} = \frac{\partial}{\partial r}\left[(1-\nu)\nabla^2\Phi - \frac{\partial^2 \Phi}{\partial z^2}\right]$$

The general solution of the problems showing irrotational deformation must be considered as the sum of the contribution from the Lame's strain potential and the one from the Loves' strain potential.

For the axi-symmetric spherical problems too, we consider the general solution as the sum of the solution given by the Lame's strain potential $\varphi(R,\phi)$ and that by the Love's strain potential $\Phi(R,\phi)$. In terms of the displacement, this can be written from Eqs. (1.34) and (1.36) as

$$2\mu \boldsymbol{u} = \hat{\nabla}\varphi + 2(1-\nu)(\hat{\nabla}^2\Phi)\boldsymbol{i}_z - \hat{\nabla}(\hat{\nabla}\cdot\Phi\boldsymbol{i}_z) \tag{1.39}$$

where $\hat{\nabla}$ denotes ∇ with $\partial/\partial\theta = 0$ in the spherical coordinates. Decomposing $\boldsymbol{i}_z = \boldsymbol{i}_R\cos\phi - \boldsymbol{i}_\phi\sin\phi$, we can show that each component can be written as

$$2\mu u_R = \frac{\partial \varphi}{\partial R} + \cos\phi\left[2(1-\nu)\hat{\nabla}^2\Phi - \frac{\partial^2\Phi}{\partial R^2}\right] + \frac{\sin\phi}{R}\frac{\partial}{\partial\phi}\left(\frac{\partial\Phi}{\partial R} - \frac{\Phi}{R}\right) \tag{1.40a}$$

$$2\mu u_\phi = \frac{1}{R}\frac{\partial\varphi}{\partial\phi} + \sin\phi\left[-2(1-\nu)\hat{\nabla}^2\Phi + \frac{1}{R}\frac{\partial\Phi}{\partial R} + \frac{1}{R^2}\frac{\partial^2\Phi}{\partial\phi^2}\right] + \frac{\cos\phi}{R}\frac{\partial}{\partial\phi}\left(-\frac{\partial\Phi}{\partial R} + \frac{\Phi}{R}\right) \tag{1.40b}$$

Note that the first terms on the right-hand side correspond to the Lame's potential solutions. Now, substitute these again into the displacement-strain relations and then the strains into the stress-strain relations give the followings.

$$\sigma_R = \frac{\partial^2\varphi}{\partial R^2} + \cos\phi\frac{\partial}{\partial R}\left[(2-\nu)\hat{\nabla}^2\Phi - \frac{\partial^2\Phi}{\partial R^2}\right] + \frac{\sin\phi}{R}\frac{\partial}{\partial\phi}\left[-\nu\hat{\nabla}^2\Phi + \frac{\partial^2\Phi}{\partial R^2} - \frac{2}{R}\frac{\partial\Phi}{\partial R} + \frac{2\Phi}{R^2}\right] \tag{1.41a}$$

$$\sigma_\theta = \left(\cos\phi\frac{\partial}{\partial R} - \frac{\sin\phi}{R}\frac{\partial}{\partial\phi}\right)\cdot\left[-(1-\nu)\hat{\nabla}^2\Phi + \frac{\partial^2\Phi}{\partial R^2} + \frac{1}{R}\frac{\partial\Phi}{\partial R} + \frac{1}{R^2}\frac{\partial^2\Phi}{\partial\phi^2}\right] \tag{1.41b}$$

$$\sigma_\phi = \frac{1}{R}\frac{\partial\varphi}{\partial R} + \frac{1}{R^2}\frac{\partial^2\varphi}{\partial\phi^2} + \cos\phi\frac{\partial}{\partial R}\left[\nu\hat{\nabla}^2\Phi - \frac{1}{R}\frac{\partial\Phi}{\partial R} - \frac{1}{R^2}\frac{\partial^2\Phi}{\partial\phi^2}\right] + \frac{\sin\phi}{R}\frac{\partial}{\partial\phi}\left[-(2-\nu)\hat{\nabla}^2\Phi + \frac{3}{R}\frac{\partial\Phi}{\partial R} - \frac{2\Phi}{R^2} + \frac{1}{R^2}\frac{\partial^2\Phi}{\partial\phi^2}\right] \tag{1.41c}$$

$$\tau_{R\phi}=\frac{\partial^2(\varphi/R)}{\partial R\partial\phi}+\frac{\cos\phi}{R}\frac{\partial}{\partial\phi}\left[(1-\nu)\hat{\nabla}^2\Phi-\frac{\partial^2\Phi}{\partial R^2}+\frac{2}{R}\frac{\partial\Phi}{\partial R}-\frac{2\Phi}{R^2}\right] \tag{1.41d}$$
$$+\sin\phi\frac{\partial}{\partial R}\left[-(1-\nu)\hat{\nabla}^2\Phi+\frac{1}{R}\frac{\partial\Phi}{\partial R}+\frac{1}{R^2}\frac{\partial^2\Phi}{\partial\phi^2}\right]$$

To obtain the Lame's strain potential φ from the solution of the Laplace equation

$$\hat{\nabla}^2\varphi=\frac{1}{R^2}\frac{\partial}{\partial R}\left(R^2\frac{\partial\varphi}{\partial R}\right)+\frac{1}{R^2\sin\phi}\frac{\partial}{\partial\phi}\left(\sin\phi\frac{\partial\varphi}{\partial\phi}\right)=0$$

we first replace the variable ϕ with ξ by using $\xi=\cos\phi$. Then we assume a separable form, $\varphi=R^nG_n(\phi)$, where n denotes the modal number. Substituting this into the above equation leads to the Legendre equation

$$(1-\xi^2)G_n''-2\xi G_n'+n(n+1)G_n=0$$

The solution to this equation, for integer n, is known as Legendre polynomials. Useful solutions are the first kinds, $P_n(\xi)$;

$$P_0(\xi)=1,\quad P_1(\xi)=\xi,\quad P_2(\xi)=\frac{1}{2}(3\xi^2-1),\quad P_3(\xi)=\frac{1}{2}(5\xi^3-3\xi)$$
$$P_4(\xi)=\frac{1}{8}(35\xi^4-30\xi^2+3),\quad P_5(\xi)=\frac{1}{8}(63\xi^5-70\xi^3+15\xi),\ \ldots \tag{1.42}$$

It can be shown that the general solution of φ takes the linear combination of the terms in the form $R^nP_n(\xi)$ with $n=0,1,2,\cdots$, i.e.,

$$\left\{1,\ RP_1(\xi),\ R^2P_2(\xi),\ R^3P_3(\xi),\ \cdots\right\} \tag{1.43}$$

in the ascending series of R and those in the form $R^{-n-1}P_n(\xi)$ with $n=0,1,2,\cdots$, i.e.,

$$\left\{R^{-1},\ R^{-2}P_1(\xi),\ R^{-3}P_2(\xi),\ R^{-4}P_3(\xi),\ \cdots\right\} \tag{1.44}$$

in the descending series of R. Then Φ takes linear combination of terms in the form $R^{n+2}P_n(\xi)$ and those in the form $R^{-n+1}P_n(\xi)$ with $n=0,1,2,\cdots$. However, since the Love's strain potential solution also carries unknown constants, those constants for φ and Φ must be determined simultaneously.

To obtain the Love's strain potential Φ, we need to solve the bi-harmonic equation

$$\hat{\nabla}^2(\hat{\nabla}^2\Phi)=0$$

In many cases, however, we do not have to solve this equation separately because the solutions can be derived from the solutions to the Lame's strain potential function φ. It is known that if φ is harmonic,

$$x\varphi,\ y\varphi,\ R^2\varphi$$

are bi-harmonic. So, we can choose for instance $\Phi=\text{constant}\times R^2\varphi$ as the Love's potential function.

1.7.5 Solutions for axi-symmetric problems in spherical coordinates

We now tackle a spherical cavity problem, where the medium is infinite and a spherical cavity of radius a is located at the origin of the coordinates whose surface is subjected to a stress-free environment (Fig. 1.21). A uniform tensile stress T is applied along the z-direction far away from the cavity.

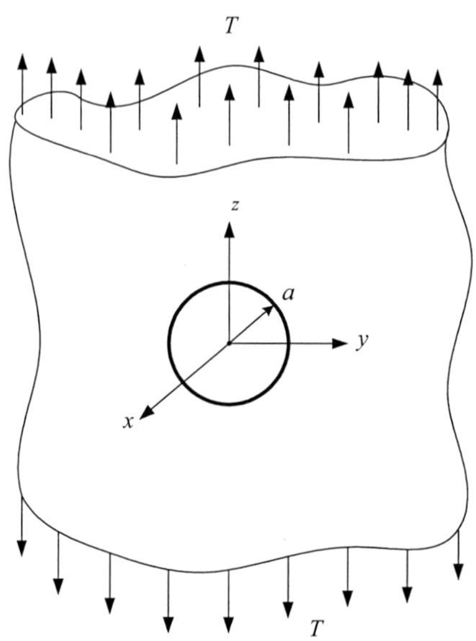

Fig. 1.21 An infinite medium having a spherical cavity subjected to a vertical tension force T.

The stress applied far away from the cavity can be written as

$$\sigma_z = T, \quad \sigma_x = \sigma_y = \tau_{xy} = \tau_{yz} = \tau_{xz} = 0$$

so that the stress tensor in the Cartesian coordinate system reads

$$\boldsymbol{A} = \begin{pmatrix} 0 & 0 & 0 \\ 0 & 0 & 0 \\ 0 & 0 & T \end{pmatrix}$$

Then we apply the transformation rule $\boldsymbol{A}' = (\boldsymbol{QA})\boldsymbol{Q}^T$ presented in Section 1.1.2 to $\boldsymbol{A}$ (stress tensor in the Cartesian coordinates, see above) to get the stress components in the spherical coordinate system $\boldsymbol{A}'$. The transformation matrix is

$$\boldsymbol{Q} = \begin{pmatrix} \sin\phi & 0 & \cos\phi \\ 0 & 1 & 0 \\ \cos\phi & 0 & -\sin\phi \end{pmatrix}$$

Then we get

$$\boldsymbol{A}' = \begin{pmatrix} \sigma_R & \tau_{R\theta} & \tau_{R\phi} \\ \tau_{R\theta} & \sigma_\theta & \tau_{\theta\phi} \\ \tau_{R\phi} & \tau_{\theta\phi} & \sigma_\phi \end{pmatrix} = \begin{pmatrix} T\cos^2\phi & 0 & -T\sin\phi\cos\phi \\ 0 & 0 & 0 \\ -T\sin\phi\cos\phi & 0 & T\sin^2\phi \end{pmatrix}$$

or

$$\sigma_R^\infty = \frac{T}{2}(1+\cos 2\phi), \quad \sigma_\phi^\infty = \frac{T}{2}(1-\cos 2\phi), \quad \tau_{R\phi}^\infty = -\frac{T}{2}\sin 2\phi$$

Obviously the far-field stress components written in the Cartesian coordinates satisfy the equilibrium equation (1.4). Therefore σ_R^∞, σ_ϕ^∞ and $\tau_{R\phi}^\infty$ must also satisfy the equilibrium equations written in the spherical coordinates. Thus, we can consider sum of the far-field solution and a perturbation to it as the general solution to this problem; $\sigma = \sigma^\infty + \tilde{\sigma}$. In order to reproduce the far-field conditions, we must require

$$\tilde{\sigma}_R, \ \tilde{\sigma}_\phi, \ \tilde{\tau}_{R\phi} \to 0 \quad \text{as } R \to \infty \tag{1.45}$$

Stress-free boundary condition for $\tilde{\sigma}$ on the cavity surface means

$$\tilde{\sigma}_R = -\frac{T}{2}(1+\cos 2\phi), \quad \tilde{\tau}_{R\phi} = \frac{T}{2}\sin 2\phi \quad \text{at } R=a \tag{1.46}$$

Therefore, we expect that the perturbation stress should be linear combination of the terms in a separable form like $f(R)g(\phi)$, where the radial function f should approach zero in the far field and the function g must take three kinds, i.e., a constant, $\cos 2\phi$ and $\sin 2\phi$. So, we try

$$\varphi = \frac{A_0}{R} + \frac{A_2}{R^3}P_2(\xi) \tag{1.47a}$$

$$\Phi = B_1 P_1(\xi) \tag{1.47b}$$

where unknown constants A_0, A_2 and B_1 are to be determined from boundary conditions (1.46). Notice that the terms on right-hand side of (1.47a) correspond to the first and third terms, respectively, in Eq. (1.44) and the right-hand side of (1.47b) is given by R^2 times the second term in (1.44). Substituting these into (1.41a) and (1.41d), we get

$$\tilde{\sigma}_R = \frac{2A_0}{R^3} + \frac{A_2}{R^5}(9\cos 2\phi + 3) + \frac{B_1}{R^3}[(5-\nu)\cos 2\phi + 3(1-\nu)] \tag{1.48a}$$

$$\tilde{\tau}_{R\phi} = \frac{6A_2}{R^5}\sin 2\phi + \frac{B_1}{R^3}(1+\nu)\sin 2\phi \tag{1.48b}$$

Applying the boundary conditions (1.46) results in the three algebraic equations, which can be solved to yield

$$A_0 = \frac{a^3(1-5\nu)T}{2(7-5\nu)}, \quad A_2 = \frac{a^5 T}{(7-5\nu)}, \quad B_1 = \frac{-5a^3 T}{2(7-5\nu)}$$

The final solutions can be obtained by substituting these constants back to the equations for the stress, e.g., Eqs. (1.48a) and (1.48b), and then adding the far-field solutions.

Homework

Derive the final solutions of σ_R, σ_ϕ, σ_θ and $\tau_{R\phi}$. Evaluate the stresses on the cavity surface and show that $\sigma_R = 0$ indeed. Draw the maximum value of σ_ϕ and σ_θ on the cavity surface and compare your result with the corresponding ones given in the cylindrical cavity treated in Section 1.7.3. Investigate how σ_R changes along the vertical axis, $\phi = 0$.

Chapter 2. Laminar Flow: Governing Equations

2.1 Basics

2.1.1 Statistical and continuum methods

There are basically two methods to derive the governing equations for the fluid flow: statistical and continuum methods. In the statistical method as a kind of molecular approach, the fluid is treated as consisting of molecules whose motion is governed by the laws of dynamics. In the continuum method, on the other hand, individual molecular motions are ignored and the fluid is assumed to consist of continuous matter. The continuum method requires that the mean free path of the molecules should be very small compared with the smallest physical-length scale of the flow field. Hence, the fluid properties represent average characteristics of its molecular structure. For instance, consider a small volume of fluid $\triangle V$ containing a large number of molecules whose total mass is $\triangle m$ (Fig. 2.1). Here, the density ρ is defined as

$$\rho \equiv \lim_{\triangle V \to \delta V} \frac{\triangle m}{\triangle V}$$

where δV is the smallest volume of $\triangle V$, for which the density variation is smooth, and thus the limit of the continuum hypothesis. Therefore, the former method is appropriate for analyzing light or rarefied gases, whereas the latter one is for analyzing heavy gases or liquids. In this part, we have selected the continuum method to derive the governing equations for the fluid flow.

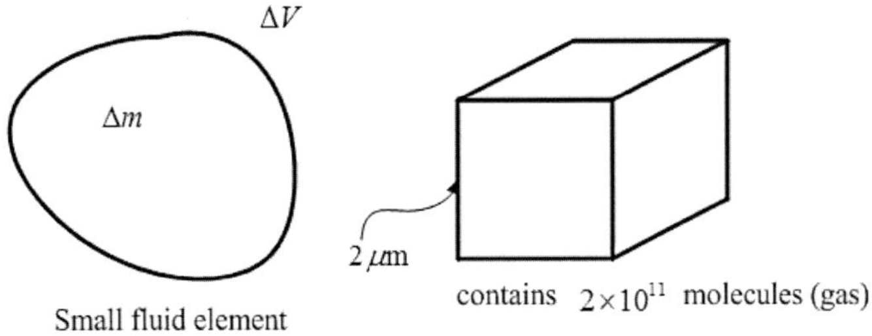

Fig. 2.1 Sketch showing a small mass inside a small volume (left), and a cube of side length 2μm containing very large number of molecules even for a typical gas (right).

2.1.2 Lagrangian and Eulerian coordinates

There are also two basic approaches to describe the fluid flow: Lagrangian and Eulerian approaches. In the Lagrangian approach, attention is fixed on a particular mass of fluid as it flows (Fig. 2.2). Here, the independent variables are (x_0, y_0, z_0) and t, where (x_0, y_0, z_0) is the coordinate through which a specified fluid element passed at a time t_0. That is, the coordinate (x_0, y_0, z_0) identifies which fluid element is being considered, and the time t identifies its instantaneous location.

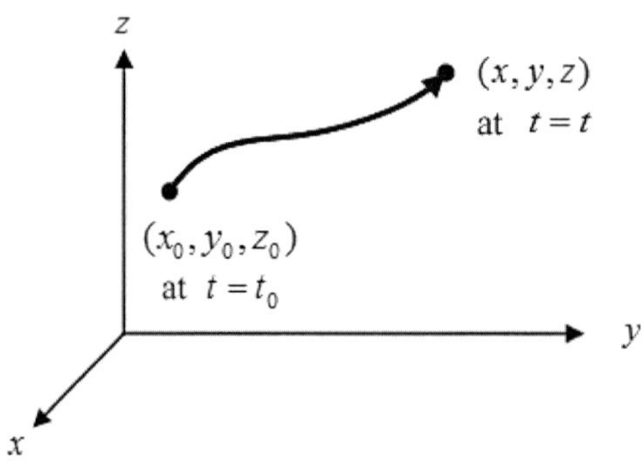

Fig. 2.2 Motion of a particle described with Cartesian coordinates and time.

In the Eulerian approach, on the other hand, attention is focussed on the fixed control volume at the location (x, y, z) through which the fluid passes (Fig. 2.3). Here, the independent variables are (x, y, z) and t, where (x, y, z) is the coordinate which identifies the control volume.

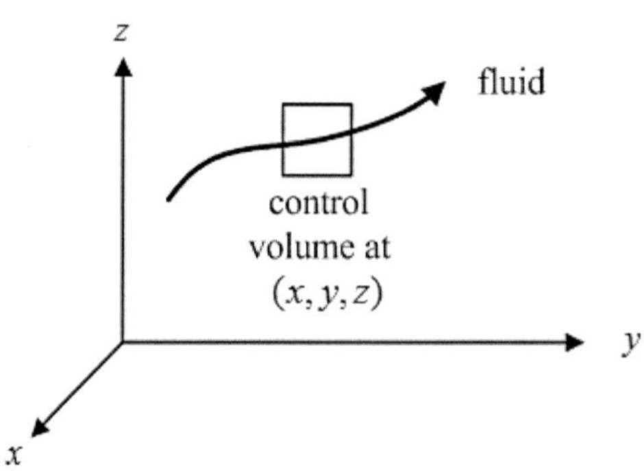

Fig. 2.3 A control volume fixed in a space for observing the particle's motion.

It is usual that the Lagrangian approach is introduced when the physical laws are applied and that the Eulerian approach is used when the problems are to be solved and analyzed. The relation between the two approaches is given by the concept of material derivative and the Reynolds transport theorem.

2.1.3 Material derivative

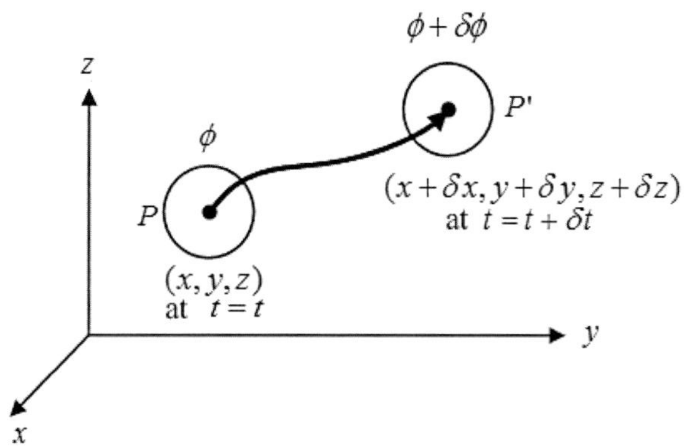

Fig. 2.4 Concept diagram illustrating that the variable ϕ associated with a fluid particle can vary in time while the particle moves in the space.

Let ϕ be any field variable defined on the Eulerian coordinate, that is $\phi=\phi(x,y,z,t)$. As shown in Fig. 2.4, a specified fluid element located at the position (x,y,z) at the time t is assumed to move by an amount $(\delta x,\delta y,\delta z)$ for a very short time interval δt. Simultaneously, the field variable which the element is supposed to carry will also change by an amount $\delta\phi$. During the time δt the change of the field variable $\delta\phi$ in the Lagrangian framework can be calculated to be

$$\delta\phi=\frac{\partial\phi}{\partial t}\delta t+\frac{\partial\phi}{\partial x}\delta x+\frac{\partial\phi}{\partial y}\delta y+\frac{\partial\phi}{\partial z}\delta z$$

because (x,y,z) is no longer an independent variable but a function of t. Here, $(\delta x,\delta y,\delta z)$ is the coordinate difference between the points P and P'. The change rate becomes

$$\frac{\delta\phi}{\delta t}=\frac{\partial\phi}{\partial t}+\frac{\delta x}{\delta t}\frac{\partial\phi}{\partial x}+\frac{\delta y}{\delta t}\frac{\partial\phi}{\partial y}+\frac{\delta z}{\delta t}\frac{\partial\phi}{\partial z}$$

As δt tends to zero, the rate of change in ϕ as observed in the Lagrangian framework becomes

$$\frac{D\phi}{Dt}=\frac{\partial\phi}{\partial t}+u\frac{\partial\phi}{\partial x}+v\frac{\partial\phi}{\partial y}+w\frac{\partial\phi}{\partial z}=\frac{\partial\phi}{\partial t}+(\mathrm{u}\cdot\nabla)\phi \qquad (2.1)$$

which is called the material derivative. In the tensor form, the above equation can be rewritten as

$$\frac{D\phi}{Dt}=\frac{\partial\phi}{\partial t}+u_k\frac{\partial\phi}{\partial x_k} \tag{2.2}$$

Here, $D\phi/Dt$ represents the total change in the quantity ϕ as seen by an observer who is following the fluid and is watching a particular mass of the fluid. The right-hand side of Eq. (2.2), on the other hand, denotes the total change in the quantity ϕ expressed in the Eulerian framework. $\partial\phi/\partial t$ is the rate of change in ϕ measured at the fixed point P, indicating the temporal change. $u_k(\partial\phi/\partial x_k)$ measures the rate of change in ϕ following the streamline at a fixed instant of time t, indicating the spatial change.

2.2 Reynolds Transport Theorem

The Reynolds transport theorem permits transformation from the material derivative of a volume integral to the volume integral of an Eulerian derivative. Consider a field variable ϕ in the Lagrangian coordinates, that is $\phi=\phi(x_0,y_0,z_0,t)$ (Fig. 2.5).

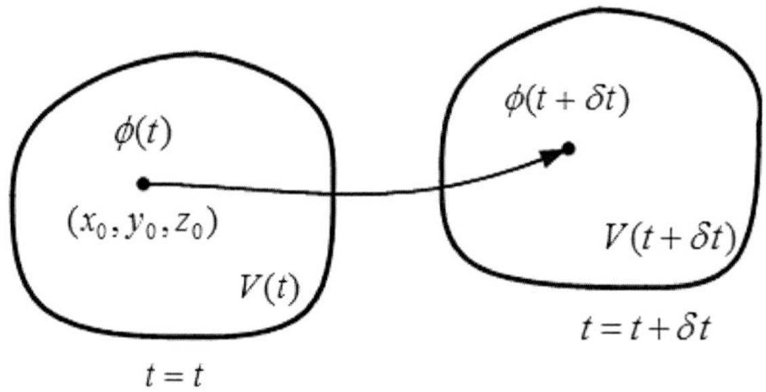

Fig. 2.5 Schematic illustrating that an identified fluid blob can change its volume as well as the variable ϕ carried by the fluid while moving.

Since the fluid blob carrying the variable ϕ is identified, the spatial coordinates (x_0,y_0,z_0) are constants; that is, $\phi=\phi(t)$. Consider the material derivative of the volume integral of ϕ:

$$\begin{aligned}\frac{D}{Dt}\int_{V(t)}\phi dV &= \lim_{\delta t\to 0}\frac{1}{\delta t}\left[\int_{V(t+\delta t)}\phi(t+\delta t)dV-\int_{V(t)}\phi(t)dV\right]\\ &= \lim_{\delta t\to 0}\frac{1}{\delta t}\left[\int_{V(t+\delta t)}\phi(t+\delta t)dV-\int_{V(t)}\phi(t+\delta t)dV\right.\\ &\quad\left.+\int_{V(t)}\phi(t+\delta t)dV-\int_{V(t)}\phi(t)dV\right]\end{aligned}$$

$$= \underbrace{\lim_{\delta t \to 0} \frac{1}{\delta t} \left[\int_{V(t+\delta t) - V(t)} \phi(t+\delta t) dV \right]}_{I}$$

$$+ \underbrace{\lim_{\delta t \to 0} \frac{1}{\delta t} \left\{ \int_{V(t)} [\phi(t+\delta t) - \phi(t)] dV \right\}}_{II} \tag{2.3}$$

As the control volume V is assumed to be finite and δt approaches zero, the volumes $V(t+\delta t)$ and $V(t)$ are almost the same except for a small difference.

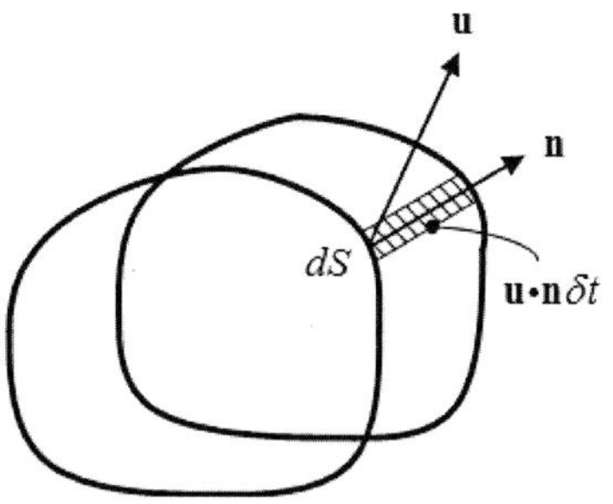

Fig. 2.6 Schematic illustrating how the infinitesimal volume dV can be represented in terms of the velocity and the surface area.

The small volume difference dV within the domain $V(t+\delta t) - V(t)$ can be constructed by the hatched column in Fig. 2.6; $dV = (\mathrm{u} \cdot \mathrm{n})\delta t\, dS$. Therefore,

$$\mathrm{I} = \lim_{\delta t \to 0} \frac{1}{\delta t} \int_{S(t)} \phi(t+\delta t)(\mathrm{u} \cdot \mathrm{n}) \delta t dS = \int_{S(t)} \phi(t)(\mathrm{u} \cdot \mathrm{n}) dS$$

where $S(t)$ is the surface surrounding the volume $V(t)$. On the other hand,

$$\mathrm{II} = \lim_{\delta t \to 0} \frac{1}{\delta t} \int_{V(t)} [\phi(t+\delta t) - \phi(t)] dV$$

Since the volume for this integral is fixed, this corresponds to the temporal rate of change in the Eulerian concept, that is,

$$\mathrm{II} = \int_{V(t)} \frac{\partial \phi}{\partial t} dV$$

Thus, we have

$$\frac{D}{Dt} \int_V \phi(t) dV = \int_V \frac{\partial \phi}{\partial t} dV + \int_S \phi(\mathrm{u} \cdot \mathrm{n}) dS \tag{2.4}$$

The relation (2.4) is referred to as the Reynolds transport theorem.

Now applying the Gauss theorem gives

$$\int_S \phi(\mathrm{u} \cdot \mathrm{n}) dS = \int_V \nabla (\phi \mathrm{u}) dV$$

Thus,

$$\frac{D}{Dt}\int_V \phi(t)dV = \int_V \left[\frac{\partial \phi}{\partial t} + \nabla(\phi \mathbf{u})\right] dV$$

$$= \int_V \left[\frac{\partial \phi}{\partial t} + \frac{\partial}{\partial x_k}(\phi u_k)\right] dV \tag{2.5}$$

Exercise

Consider the following two-dimensional flow and temperature fields (Fig. 2.7):

Flow field: $u = x\,,\ v = -y$

Temperature: $T = x + t$

Domain at $t = 0$: $1 \le x \le 2$ and $0 \le y \le 1$

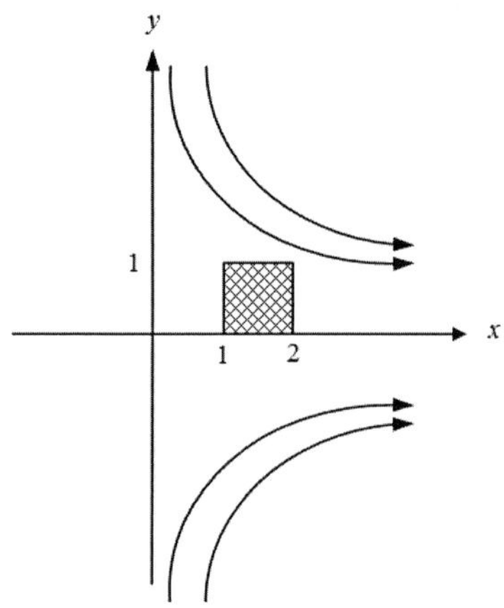

Fig. 2.7 Two-dimensional flow field and an initial fluid blob of square shape.

At $t=0$: $\int_{V(0)} T dV = \int_0^1 \int_1^2 x\, dx dy = 1.5$

At $t = \Delta t = 0.1$: $\int_{V(0.1)} T dV = \int_0^{0.9} \int_{1.1}^{2.2} (x + 0.1)\, dx dy = 1.7325$

$$\frac{D}{Dt}\int_{V(0)} T dV \cong \frac{1}{\Delta t}\left[\int_{V(0.1)} T dV - \int_{V(0)} T dV\right] = \frac{1.7325 - 1.5}{0.1} = 2.325$$

On the other hand,

$$\frac{\partial T}{\partial t} + \mathbf{u}\cdot\nabla T = 1 + u\frac{\partial T}{\partial x} + v\frac{\partial T}{\partial y} = 1 + x$$

$$\int_{V(0)}\left(\frac{\partial T}{\partial t} + \mathbf{u}\cdot\nabla T\right) dV = \int_0^1 \int_1^2 (1+x)\, dx dy = \int_0^1 \left[x + \frac{1}{2}x^2\right]_1^2 dy = 2.5$$

Therefore, we can roughly conclude that

$$\frac{D}{Dt}\int T dV \cong \int \left(\frac{\partial T}{\partial t} + \mathbf{u}\cdot\nabla T\right) dV.$$

Homework

Answer to the following questions:

1) Do the same work as the above exercise for $\Delta t = 0.01$ and 0.001. What is your conclusion?

2) Do the same work for u and v instead of T.

2.3 Conservation Laws

2.3.1 Conservation of mass: continuity equation

The mass contained in an arbitrary volume V must be unchanged while it flows:

$$\frac{D}{Dt}\int_V \rho dV = 0 \quad \text{or} \quad \int_V \left[\frac{\partial \rho}{\partial t} + \frac{\partial}{\partial x_k}(\rho u_k)\right] dV = 0$$

Note that the above equation is derived by replacing the field variable ϕ with the density ρ in Eq. (2.5). Since the volume V is arbitrarily chosen, we must require

$$\frac{\partial \rho}{\partial t} + \frac{\partial}{\partial x_k}(\rho u_k) = 0$$

$$\text{or} \quad \frac{\partial \rho}{\partial t} + u_k \frac{\partial \rho}{\partial x_k} + \rho \frac{\partial u_k}{\partial x_k} = \frac{D\rho}{Dt} + \rho \frac{\partial u_k}{\partial x_k} = 0 \tag{2.6}$$

The derived differential equation is called the continuity equation. For the incompressible fluid ($D\rho/Dt = 0$), it holds that

$$\frac{\partial u_k}{\partial x_k} = 0 \tag{2.7}$$

2.3.2 Conservation of momentum: Navier-Stokes equation

Applying the Newton's second law of motion to a fluid element or a fluid system, we have

$$\sum_{\text{on system}} \mathbf{F} = \underbrace{\frac{d}{dt}(m\mathbf{u})}_{\text{time rate of momentum change}}$$

The external force acting on the system comprises of the surface force and the body force (Fig. 2.8) as follows:

$$\sum_{\text{on system}} \mathbf{F} = \sum_{\text{on surface}} \mathbf{F} + \sum_{\text{on body}} \mathbf{F}$$

Here, the surface force includes the pressure force and the shear force, while the body force includes the gravitational force or the electromagnetic force. Let

$$\sum_{\text{on surface}} \mathrm{F} = \int_S \mathrm{P}\, dS$$

where P represents the surface force vector per unit area. Similarly,

$$\sum_{\text{on body}} \mathrm{F} = \int_V \rho \mathrm{f}\, dV$$

where f is the body force vector per unit mass. Note that

$$\frac{d}{dt}(m\mathrm{u}) = \frac{D}{Dt}\int_V \rho \mathrm{u}\, dV$$

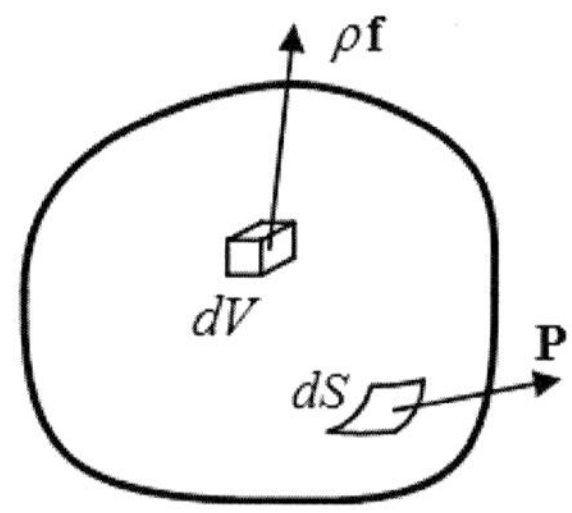

Fig. 2.8 Body force acting on fluid volume dV and surface force acting on a fluid surface dS.

Therefore,

$$\frac{D}{Dt}\int_V \rho \mathrm{u}\, dV = \int_S \mathrm{P}\, dS + \int_V \rho \mathrm{f}\, dV \tag{2.8}$$

Here, it will be convenient to use a hexahedron.

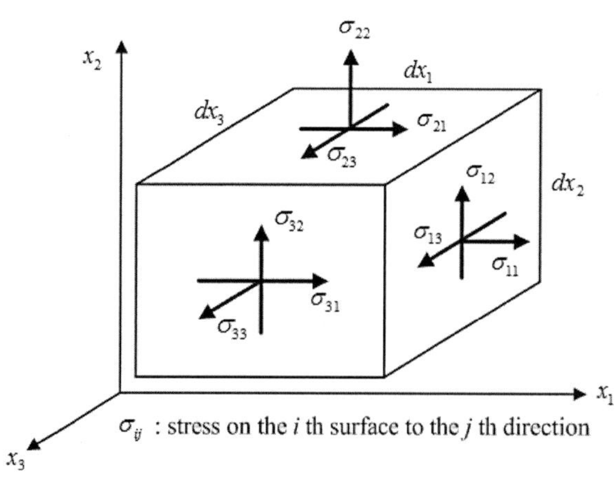

Fig. 2.9 Normal (σ) and tangential (τ) stresses acting on three frontal faces of an infinitesimal 3-D element in the fluid.

As shown in Fig. 2.9, the surface force (per unit area) on the three front faces can be written as

$$\mathrm{P}_f = \begin{cases} \sigma_{11}\mathbf{i} + \sigma_{12}\mathbf{j} + \sigma_{13}\mathbf{k} & \text{on surface 1} \\ \sigma_{21}\mathbf{i} + \sigma_{22}\mathbf{j} + \sigma_{23}\mathbf{k} & \text{on surface 2} \\ \sigma_{31}\mathbf{i} + \sigma_{32}\mathbf{j} + \sigma_{33}\mathbf{k} & \text{on surface 3} \end{cases}$$

where σ_{ij} represents the stress on the ith surface to the jth direction. On the rear 3 faces, σ_{ij} will be assumed to be smaller than that of the front face as much as $\dfrac{\partial \sigma_{ij}}{\partial x_i} dx_i$. For instance,

$$\sigma_{11r} = \sigma_{11f} - \frac{\partial \sigma_{11}}{\partial x_1} dx_1$$

Thus

$$\begin{aligned} \int_{\text{total}} \mathrm{P}\, dS &= \int_{\text{front}} \mathrm{P}\, dS - \int_{\text{rear}} \mathrm{P}\, dS \\ &= \int_V \frac{\partial}{\partial x_1}(\sigma_{11}\mathbf{i} + \sigma_{12}\mathbf{j} + \sigma_{13}\mathbf{k})\, dx_1 dx_2 dx_3 \\ &\quad + \int_V \frac{\partial}{\partial x_2}(\sigma_{21}\mathbf{i} + \sigma_{22}\mathbf{j} + \sigma_{23}\mathbf{k})\, dx_1 dx_2 dx_3 \\ &\quad + \int_V \frac{\partial}{\partial x_3}(\sigma_{31}\mathbf{i} + \sigma_{32}\mathbf{j} + \sigma_{33}\mathbf{k})\, dx_1 dx_2 dx_3 \\ &= \int_V \frac{\partial}{\partial x_k}(\sigma_{k1}\mathbf{i} + \sigma_{k2}\mathbf{j} + \sigma_{k3}\mathbf{k})\, dx_1 dx_2 dx_3 \end{aligned}$$

In fact, we have changed the surface integral to the volume integral.

Next, using $\mathbf{u} = u_1\mathbf{i} + u_2\mathbf{j} + u_3\mathbf{k}$ we can write

$$\frac{D}{Dt}\int_V \rho \mathbf{u}\, dV = \frac{D}{Dt}\int_V \rho u_i\, dV$$

The Reynolds transport theorem then applies to yield

$$\begin{aligned} \frac{D}{Dt}\int_V \rho \mathbf{u}\, dV &= \frac{D}{Dt}\int_V \rho u_i\, dV \\ &= \int_V \left[\frac{\partial}{\partial t}(\rho u_i) + \frac{\partial}{\partial x_k}(\rho u_i u_k)\right] dV \\ &= \int_V \left[\rho \frac{\partial u_i}{\partial t} + \underbrace{u_i \frac{\partial \rho}{\partial t} + u_i \frac{\partial}{\partial x_k}(\rho u_k)}_{\text{zero (continuity equation)}} + \rho u_k \frac{\partial u_i}{\partial x_k}\right] dV \end{aligned}$$

Substituting all these, we can write

$$\int_V \left[\rho \frac{\partial u_i}{\partial t} + \rho u_k \frac{\partial u_i}{\partial x_k}\right] dV = \int_V \left[\frac{\partial \sigma_{ki}}{\partial x_k} + \rho f_i\right] dV$$

or

$$\underbrace{\underbrace{\rho\frac{\partial u_i}{\partial t}}_{\text{temporal acceleration}} + \underbrace{\rho u_k\frac{\partial u_i}{\partial x_k}}_{\text{convective acceleration}}}_{\text{total acceleration}} = \underbrace{\frac{\partial \sigma_{ki}}{\partial x_k} + \rho f_i}_{\text{forces causing acceleration}} \tag{2.9}$$

2.3.3 Conservation of energy: energy equation

To a fluid element or a fluid system, let us apply the first law of thermodynamics reading "the time rate of change in the sum of the internal energy and the kinetic energy is equal to the sum of the total power done on the system and the heat flow rate added".

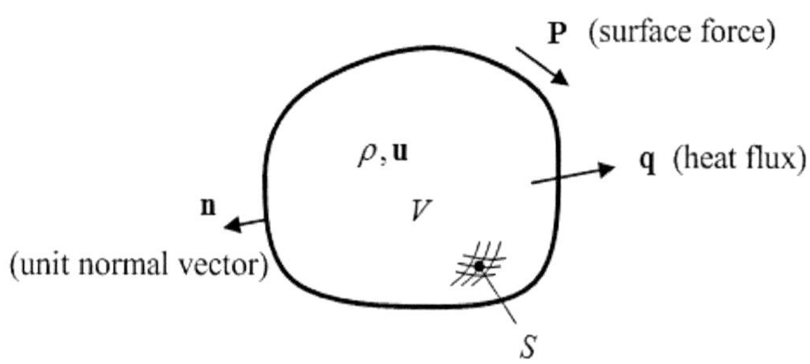

Fig. 2.10 Definition of the unit normal vector n and the heat flux vector q.

Here, the sum of the internal energy and the kinetic energy can be written as

$$\int_V \left(\rho e + \frac{1}{2}\rho \mathrm{u}\cdot\mathrm{u}\right)dV$$

where e is the internal energy per unit mass. The work per unit time (power) done by the surface force is written as

$$\int_S \mathrm{u}\cdot\mathrm{P}\,dS$$

Similarly, we can write as follows:

Power done by the body force: $\int_V \mathrm{u}\cdot\rho\mathrm{f}\,dV$

Rate of heat added through the surface: $-\int_S \mathrm{q}\cdot\mathrm{n}\,dS$

where the unit normal vector n and the heat flux vector q are directed outward as shown in Fig. 2.10. Thus, the first law of thermodynamics reads

$$\frac{D}{Dt}\int_V \left(\rho e + \frac{1}{2}\rho \mathrm{u}\cdot\mathrm{u}\right)dV = \int_S \mathrm{u}\cdot\mathrm{P}\,dS$$

$$+\int_V \mathbf{u}\cdot\rho\mathbf{f}\,dV-\int_S \mathbf{q}\cdot\mathbf{n}\,dS \tag{2.10}$$

Now,

$$\int_S \mathbf{u}\cdot\mathbf{P}\,dS=\int_V \frac{\partial}{\partial x_i}(u_j\sigma_{ij})\,dV$$

$$\int_S \mathbf{q}\cdot\mathbf{n}\,dS=\int_V \frac{\partial q_j}{\partial x_j}dV \quad \text{(from Gauss theorem)}$$

Therefore,

$$\frac{D}{Dt}\int_V\left(\rho e+\frac{1}{2}\rho u_j u_j\right)dV=\int_V \frac{\partial}{\partial x_i}(u_j\sigma_{ij})\,dV$$

$$+\int_V u_j\rho f_j\,dV-\int_V \frac{\partial q_j}{\partial x_j}dV$$

$$\frac{D}{Dt}\left(\rho e+\frac{1}{2}\rho u_j u_j\right)=\frac{\partial}{\partial x_i}(u_j\sigma_{ij})+u_j\rho f_j-\frac{\partial q_j}{\partial x_j}$$

$$\frac{\partial}{\partial t}(\rho e+\frac{1}{2}\rho u_j u_j)+\frac{\partial}{\partial x_i}\left[(\rho e+\frac{1}{2}\rho u_j u_j)u_i\right]$$

$$=\frac{\partial}{\partial x_i}(u_j\sigma_{ij})+u_j\rho f_j-\frac{\partial q_j}{\partial x_j} \tag{2.11}$$

So, we have

$$\underbrace{\underbrace{\rho\frac{\partial e}{\partial t}}_{\substack{\text{temporal}\\\text{change}\\\text{of}\\\text{energy}}}+\underbrace{\rho u_k\frac{\partial e}{\partial x_k}}_{\substack{\text{convective}\\\text{change}\\\text{of}\\\text{energy}}}}_{\substack{\text{rate of change of}\\\text{internal energy}}}=\underbrace{\underbrace{\sigma_{ij}\frac{\partial u_j}{\partial x_i}}_{\substack{\text{work done}\\\text{by the}\\\text{surface force}}}-\underbrace{\frac{\partial q_j}{\partial x_j}}_{\text{heat released}}}_{\substack{\text{cause of the}\\\text{change in the}\\\text{internal energy}}} \tag{2.12}$$

which is usually referred to as simply the "energy equation".

Homework

Prove that Eq. (2.11) can be converted into Eq. (2.12).

2.3.4 Discussion of conservation equations

Up to now, we have obtained a total of 7 equations:

Continuity equation	1	(scalar)
Momentum equations	3	(vector)
Energy equation	1	(scalar)
State equations (thermodynamics)	2	(scalar)

But, these equations include a total of 19 unknowns:

u_i	3	(vector)
q_i	3	(vector)
σ_{ij}	9	(tensor)
ρ, e, p, T	4	(scalar)

In order to obtain a complete set of equations, the stress tensor σ_{ij} and the heat-flux vector q_i must be further specified, which leads to the so-called "constitutive equations", where the stress tensor is related to the deformation-rate tensor and the heat-flux vector is related to the temperature gradients. We have experienced similar relation in the Elasticity, where the stress tensor is related to the deformation tensor.

2.4 Governing Equations

2.4.1 Rotation and shear rate (strain rate)

Consider a two dimensional flow where the deformation of a rectangle $(\delta x) \times (\delta y)$, shown in Fig. 2.11, is to be discussed.

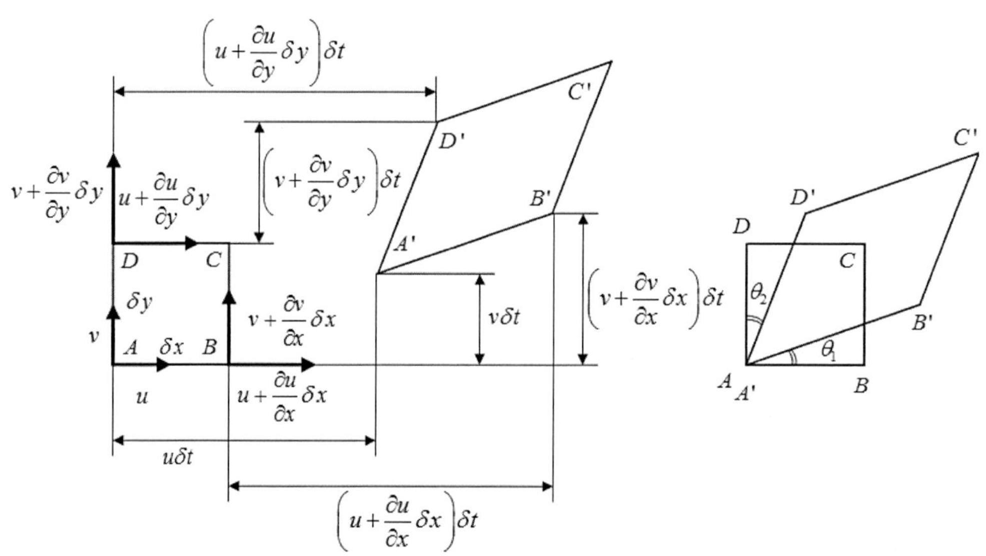

Fig. 2.11 The initial fluid element ABCD and the element A'B'C'D' displaced as well as deformed after a small time interval δt.

The rectangle $ABCD$ at $t=0$ moves and deforms to $A'B'C'D'$ at $t=\delta t$. Superimpose $ABCD$ and $A'B'C'D'$. Define

$$\varepsilon_{xx}=\frac{1}{\delta t}\left[\frac{A'B'-AB}{AB}\right]$$

$$\varepsilon_{yy}=\frac{1}{\delta t}\left[\frac{A'D'-AD}{AD}\right]$$

$$\varepsilon_{xy}=\frac{1}{\delta t}\left[\frac{\theta_1+\theta_2}{2}\right]$$

Then

$$\varepsilon_{xx}=\frac{1}{\delta t}\left[\frac{\delta x+\left(u+\dfrac{\partial u}{\partial x}\delta x\right)\delta t-u\delta t-\delta x}{\delta x}\right]=\frac{\partial u}{\partial x}$$

$$\varepsilon_{yy}=\frac{1}{\delta t}\left[\frac{\delta y+\left(v+\dfrac{\partial v}{\partial y}\delta y\right)\delta t-v\delta t-\delta y}{\delta y}\right]=\frac{\partial v}{\partial y}$$

$$\begin{aligned}\varepsilon_{xy}&=\frac{1}{2\delta t}\left[\frac{\left(v+\dfrac{\partial v}{\partial x}\delta x\right)\delta t-v\delta t}{\delta x}+\frac{\left(u+\dfrac{\partial u}{\partial y}\delta y\right)\delta t-u\delta t}{\delta y}\right]\\&=\frac{1}{2}\left(\frac{\partial u}{\partial y}+\frac{\partial v}{\partial x}\right)\end{aligned}$$

In general, the strain-rate tensor ε_{ij} can be written as

$$\varepsilon_{ij}=\frac{1}{2}\left(\frac{\partial u_i}{\partial x_j}+\frac{\partial u_j}{\partial x_i}\right)\qquad (i,\ j=1,2,3)\tag{2.13}$$

2.4.2 Constitutive equations and viscosity

For viscous flow of compressible Newtonian fluid, the relation between the stress tensor σ_{ij} and the strain rate tensor ε_{ij} under certain assumptions is found to be

$$\sigma_{ij}=-p\delta_{ij}-\frac{2}{3}\mu\delta_{ij}\varepsilon_{kk}+2\mu\varepsilon_{ij}\tag{2.14}$$

where p is the static pressure and μ is the dynamic viscosity of the fluid. This is in fact composed of 6 independent equations because $\sigma_{ij}=\sigma_{ji}$ due to the symmetric property. Note that the right-hand side of Eq. (2.14) contains only four existing unknowns, u_i and p.

The Fourier's law of heat conduction states that the heat flux by conduction is proportional to the negative of temperature gradient:

$$q_i=-k\frac{\partial T}{\partial x_i}\tag{2.15}$$

where k is the thermal conductivity of the fluid. We have also converted the three unknowns q_i to the expressions in terms of one existing unknown T.

2.4.3 Navier-Stokes equations

In order to substitute (2.14) into (2.9), we write $\partial\sigma_{ki}/\partial x_k$ as follows:

$$\frac{\partial\sigma_{ki}}{\partial x_k} = \frac{\partial}{\partial x_k}\left[-p\delta_{ki} - \frac{2}{3}\mu\delta_{ki}\left(\frac{\partial u_m}{\partial x_m}\right) + \mu\left(\frac{\partial u_k}{\partial x_i} + \frac{\partial u_i}{\partial x_k}\right)\right]$$

$$= -\frac{\partial p}{\partial x_i} - \frac{2}{3}\mu\frac{\partial}{\partial x_i}\left(\frac{\partial u_m}{\partial x_m}\right) + \mu\frac{\partial}{\partial x_k}\left(\frac{\partial u_k}{\partial x_i} + \frac{\partial u_i}{\partial x_k}\right) \tag{2.16}$$

where μ is assumed to be constant. Then, Eq. (2.9) becomes

$$\rho\frac{\partial u_i}{\partial t} + \rho u_k\frac{\partial u_i}{\partial x_k} = -\frac{\partial p}{\partial x_i} - \frac{2}{3}\mu\frac{\partial}{\partial x_i}\left(\frac{\partial u_m}{\partial x_m}\right) + \mu\frac{\partial}{\partial x_k}\left(\frac{\partial u_k}{\partial x_i} + \frac{\partial u_i}{\partial x_k}\right) + \rho f_i \tag{2.17a}$$

or

$$\frac{\partial u_i}{\partial t} + u_k\frac{\partial u_i}{\partial x_k} = -\frac{1}{\rho}\frac{\partial p}{\partial x_i} - \frac{2}{3}\nu\frac{\partial}{\partial x_i}\left(\frac{\partial u_m}{\partial x_m}\right) + \nu\frac{\partial}{\partial x_k}\left(\frac{\partial u_k}{\partial x_i} + \frac{\partial u_i}{\partial x_k}\right) + f_i \tag{2.17b}$$

where $\nu = \mu/\rho$ is the kinematic viscosity. The derived equation (2.17) is usually referred to simply as the "Navier-Stokes equation".

For incompressible flow, this reduces to

$$\rho\frac{\partial u_i}{\partial t} + \rho u_k\frac{\partial u_i}{\partial x_k} = -\frac{\partial p}{\partial x_i} + \mu\frac{\partial^2 u_i}{\partial x_k\partial x_k} + \rho f_i \tag{2.18}$$

Homework

Confirm all the equations (2.16-18) derived in this subsection

2.4.4 Energy equations

The term $\sigma_{ij}(\partial u_j/\partial x_i)$ appearing in (2.12) becomes

$$\sigma_{ij}\frac{\partial u_j}{\partial x_i} = \left[-p\delta_{ij} - \frac{2}{3}\mu\delta_{ij}\left(\frac{\partial u_m}{\partial x_m}\right) + \mu\left(\frac{\partial u_i}{\partial x_j} + \frac{\partial u_j}{\partial x_i}\right)\right]\frac{\partial u_j}{\partial x_i}$$

$$= \underbrace{-p\left(\frac{\partial u_m}{\partial x_m}\right)}_{\substack{\text{reversible}\\ \text{transfer of energy}\\ \text{due to compression}}} \underbrace{-\frac{2}{3}\mu\left(\frac{\partial u_m}{\partial x_m}\right)^2 + \mu\left(\frac{\partial u_i}{\partial x_j} + \frac{\partial u_j}{\partial x_i}\right)\frac{\partial u_j}{\partial x_i}}_{\Phi\text{; irreversible dissipation function}}$$

The dissipation function Φ measures the rate at which the mechanical energy is converted to the thermal energy. This increases the internal energy of the flow. Therefore, Eq. (2.12) becomes

$$\rho\frac{\partial e}{\partial t} + \rho u_k\frac{\partial e}{\partial x_k} = -p\left(\frac{\partial u_m}{\partial x_m}\right) + \frac{\partial}{\partial x_j}\left(k\frac{\partial T}{\partial x_j}\right) + \Phi \tag{2.19}$$

This is the energy equation for a Newtonian fluid.

2.4.5 Governing equations for a Newtonian fluid

In summary, the equations governing the motion of a Newtonian fluid are

- Continuity equation

$$\frac{\partial \rho}{\partial t} + \frac{\partial}{\partial x_k}(\rho u_k) = 0 \tag{2.20}$$

- Momentum equations: Navier-Stokes equations

$$\rho\frac{\partial u_i}{\partial t} + \rho u_k\frac{\partial u_i}{\partial x_k} = -\frac{\partial p}{\partial x_i} - \frac{2}{3}\frac{\partial}{\partial x_i}\left(\mu\frac{\partial u_m}{\partial x_m}\right) + \frac{\partial}{\partial x_k}\left[\mu\left(\frac{\partial u_k}{\partial x_i} + \frac{\partial u_i}{\partial x_k}\right)\right] + \rho f_i \tag{2.21}$$

- Energy equation

$$\rho\frac{\partial e}{\partial t} + \rho u_k\frac{\partial e}{\partial x_k} = -p\frac{\partial u_m}{\partial x_m} + \frac{\partial}{\partial x_j}\left(k\frac{\partial T}{\partial x_j}\right) - \frac{2}{3}\mu\left(\frac{\partial u_m}{\partial x_m}\right)^2 + \mu\left(\frac{\partial u_i}{\partial x_j} + \frac{\partial u_j}{\partial x_i}\right)\frac{\partial u_j}{\partial x_i} \tag{2.22}$$

- Thermal equation of state

$$p = p(\rho, T) \tag{2.23}$$

- Caloric equation of state

$$e = e(\rho, T) \tag{2.24}$$

These are seven equations with seven unknowns: ρ, u_i (3 components), e, p, T.

For incompressible flow, $\rho = \text{constant}$, and we have

$$\frac{\partial u_k}{\partial x_k} = 0 \tag{2.25}$$

which simplifies the system of equations. Since ρ is given, the energy equation is decoupled from the continuity and momentum equations if the body force ρf_i is not involved. For the natural convection problem, however, the body force ρf_i is related to the temperature so that the energy equation and the continuity and momentum equations must be coupled together.

When the electromagnetic effects are important, $\mathrm{f} = (\rho_c \mathrm{E} + \mathrm{J} \times \mathrm{B})$, where ρ_c is the charge density, E is the electric field vector, J is the electric current density, and B is the magnetic field vector. These variables must be related by the physical laws known as Maxwell's equations.

When there is only the gravitational effect and the fluid is at rest,

$$0 = -\frac{\partial p}{\partial x_i} + \rho f_i$$

Then,

$$\frac{\partial p}{\partial z} = -\rho g \quad \rightarrow \quad p = -\rho g z + \text{constant}$$

In case of zero velocity, the energy equation becomes

$$\rho \frac{\partial e}{\partial t} = \frac{\partial}{\partial x_j}\left(k \frac{\partial T}{\partial x_j}\right)$$

Introducing the enthalpy $h = e + p/\rho$ and using the fact that p and ρ are constant in the stationary fluid gives

$$\rho \frac{\partial h}{\partial t} = \frac{\partial}{\partial x_j}\left(k \frac{\partial T}{\partial x_j}\right)$$

If the fluid is thermally perfect, $h = c_p T$ where c_p is the specific heat. Thus,

$$\rho c_p \frac{\partial T}{\partial t} = \frac{\partial}{\partial x_j}\left(k \frac{\partial T}{\partial x_j}\right) \tag{2.26}$$

which is referred to simply as the "equation of heat conduction".

2.4.6 Governing equations for the curvilinear orthogonal coordinate system

Assume that a new orthogonal coordinates (α, β, γ) are related to the Cartesian coordinates (x, y, z) by

$$x = x(\alpha, \beta, \gamma)$$
$$y = y(\alpha, \beta, \gamma)$$
$$z = z(\alpha, \beta, \gamma)$$

Then the scale factors h_1, h_2 and h_3 are obtained by

$$h_1 = \left[\left(\frac{\partial x}{\partial \alpha}\right)^2 + \left(\frac{\partial y}{\partial \alpha}\right)^2 + \left(\frac{\partial z}{\partial \alpha}\right)^2\right]^{1/2}$$

$$h_2 = \left[\left(\frac{\partial x}{\partial \beta}\right)^2 + \left(\frac{\partial y}{\partial \beta}\right)^2 + \left(\frac{\partial z}{\partial \beta}\right)^2\right]^{1/2}$$

$$h_3 = \left[\left(\frac{\partial x}{\partial \gamma}\right)^2 + \left(\frac{\partial y}{\partial \gamma}\right)^2 + \left(\frac{\partial z}{\partial \gamma}\right)^2\right]^{1/2}$$

giving $ds^2 = dx^2 + dy^2 + dz^2 = h_1^2\, d\alpha^2 + h_2^2\, d\beta^2 + h_3^2\, d\gamma^2$. The scale factor, for instance h_1, can be understood in the Cartesian coordinates simply as the ratio of the Euclidean distance between two consecutive lines of constant α and difference in the α-coordinates, $\Delta\alpha$, i.e., $h_1 = \Delta s_1/\Delta\alpha$, as shown in Fig. 2.12.

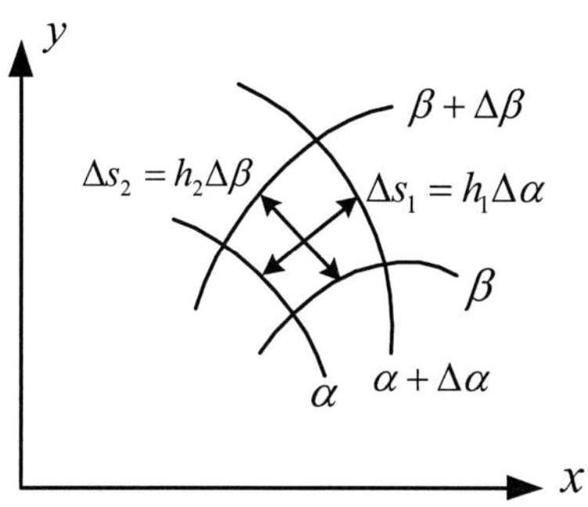

Fig. 2.12 Lines of constant values of transformed coordinates α and β, and the Euclidean distances Δs_1 and Δs_2 between them.

As an example, consider the cylindrical coordinates:

$$(r, \theta, z) = (\alpha, \beta, \gamma)$$

$$x = r\cos\theta$$

$$y = r\sin\theta$$

$$z = z$$

$$h_1 = \sqrt{\cos^2\theta + \sin^2\theta + 0} = 1$$

$$h_2 = \sqrt{(-r\sin\theta)^2 + (r\cos\theta)^2 + 0} = r$$

$$h_3 = \sqrt{0+0+1} = 1$$

Let u_1, u_2 and u_3 be the velocity components along the x-, y- and z-directions, respectively, and u, v and w are along the α-, β- and γ-directions, respectively. Then the continuity equation can be written as follows:

$$\frac{\partial \rho}{\partial t} + \frac{1}{h_1 h_2 h_3}\left[\frac{\partial}{\partial \alpha}(\rho u h_2 h_3) + \frac{\partial}{\partial \beta}(\rho v h_3 h_1) + \frac{\partial}{\partial \gamma}(\rho w h_1 h_2)\right] = 0 \qquad (2.27)$$

The Navier-Stokes equations can also be written in the same way, although their form is somewhat complicated.

Homework

Referring to the generalized continuity and Navier-Stokes equations in the literature that are derived for the curvilinear orthogonal coordinate system, derive the corresponding equations for the cylindrical and spherical coordinate system.

Chapter 3. Laminar Flow: Exact Solutions

In this chapter, we want to obtain some exact solutions of the Navier-Stokes equations. The equations governing the motion of the viscous incompressible Newtonian fluid with constant properties are written as follows:

Continuity equation:

$$\nabla \cdot \mathrm{u} = 0 \quad \text{or} \quad \frac{\partial u_j}{\partial x_j} = 0 \tag{3.1}$$

Momentum equation:

$$\frac{\partial \mathrm{u}}{\partial t} + (\mathrm{u} \cdot \nabla)\mathrm{u} = -\frac{1}{\rho}\nabla p + \nu \nabla^2 \mathrm{u} + \mathrm{f} \quad \text{or} \tag{3.2}$$

$$\frac{\partial u_i}{\partial t} + u_j \frac{\partial u_i}{\partial x_j} = -\frac{1}{\rho}\frac{\partial p}{\partial x_i} + \nu \frac{\partial^2 u_i}{\partial x_j \partial x_j} + f_i$$

The basic difficulty in solving these equations lies in the existence of the nonlinear convection terms involved in Eq. (3.2). In general, we have two categories for obtaining exact solutions: zero nonlinear terms and non-zero nonlinear terms. That is,

(i) $(\mathrm{u} \cdot \nabla)\mathrm{u} = 0$

- Couette flow
- Poiseuille flow
- Flow between rotating cylinders
- Stokes' first and second problems
- ......

(ii) $(\mathrm{u} \cdot \nabla)\mathrm{u} \neq 0$

- Stagnation-point flow (Hiemenz flow)
- Flow in a convergent or divergent channel
- ......

It is sometimes convenient to express the governing equations, (3.1) and (3.2), in each coordinate system while obtaining exact solutions as follows:

Cartesian $[(x, y, z)]$ coordinate system:

$$\frac{\partial u}{\partial x} + \frac{\partial v}{\partial y} + \frac{\partial w}{\partial z} = 0 \tag{3.3}$$

$$\frac{\partial u}{\partial t}+\left(u\frac{\partial u}{\partial x}+v\frac{\partial u}{\partial y}+w\frac{\partial u}{\partial z}\right) \tag{3.4a}$$
$$=-\frac{1}{\rho}\frac{\partial p}{\partial x}+\nu\left(\frac{\partial^2 u}{\partial x^2}+\frac{\partial^2 u}{\partial y^2}+\frac{\partial^2 u}{\partial z^2}\right)+f_x$$

$$\frac{\partial v}{\partial t}+\left(u\frac{\partial v}{\partial x}+v\frac{\partial v}{\partial y}+w\frac{\partial v}{\partial z}\right) \tag{3.4b}$$
$$=-\frac{1}{\rho}\frac{\partial p}{\partial y}+\nu\left(\frac{\partial^2 v}{\partial x^2}+\frac{\partial^2 v}{\partial y^2}+\frac{\partial^2 v}{\partial z^2}\right)+f_y$$

$$\frac{\partial w}{\partial t}+\left(u\frac{\partial w}{\partial x}+v\frac{\partial w}{\partial y}+w\frac{\partial w}{\partial z}\right) \tag{3.4c}$$
$$=-\frac{1}{\rho}\frac{\partial p}{\partial z}+\nu\left(\frac{\partial^2 w}{\partial x^2}+\frac{\partial^2 w}{\partial y^2}+\frac{\partial^2 w}{\partial z^2}\right)+f_z$$

Cylindrical $[(x,r,\theta)]$ coordinate system $[(x,y,z)=(x,r\cos\theta,r\sin\theta)]$:

$$\frac{\partial u_x}{\partial x}+\frac{1}{r}\frac{\partial(ru_r)}{\partial r}+\frac{1}{r}\frac{\partial u_\theta}{\partial \theta}=0 \tag{3.5}$$

$$\frac{\partial u_x}{\partial t}+\left(u_x\frac{\partial u_x}{\partial x}+u_r\frac{\partial u_x}{\partial r}+\frac{u_\theta}{r}\frac{\partial u_x}{\partial \theta}\right)=-\frac{1}{\rho}\frac{\partial p}{\partial x} \tag{3.6a}$$
$$+\nu\left\{\frac{\partial^2 u_x}{\partial x^2}+\frac{1}{r}\frac{\partial}{\partial r}\left[r\frac{\partial u_x}{\partial r}\right]+\frac{1}{r^2}\frac{\partial^2 u_x}{\partial \theta^2}\right\}+f_x$$

$$\frac{\partial u_r}{\partial t}+\left(u_x\frac{\partial u_r}{\partial x}+u_r\frac{\partial u_r}{\partial r}+\frac{u_\theta}{r}\frac{\partial u_r}{\partial \theta}-\frac{v_\theta^2}{r}\right)=-\frac{1}{\rho}\frac{\partial p}{\partial r} \tag{3.6b}$$
$$+\nu\left\{\frac{\partial^2 u_r}{\partial x^2}+\frac{\partial}{\partial r}\left[\frac{1}{r}\frac{\partial(ru_r)}{\partial r}\right]+\frac{1}{r^2}\frac{\partial^2 u_r}{\partial \theta^2}-\frac{2}{r^2}\frac{\partial u_\theta}{\partial \theta}\right\}+f_r$$

$$\frac{\partial u_\theta}{\partial t}+\left(u_x\frac{\partial u_\theta}{\partial x}+u_r\frac{\partial u_\theta}{\partial r}+\frac{u_\theta}{r}\frac{\partial u_\theta}{\partial \theta}+\frac{u_r u_\theta}{r}\right)=-\frac{1}{\rho}\frac{1}{r}\frac{\partial p}{\partial \theta} \tag{3.6c}$$
$$+\nu\left\{\frac{\partial^2 u_\theta}{\partial x^2}+\frac{\partial}{\partial r}\left[\frac{1}{r}\frac{\partial(ru_\theta)}{\partial r}\right]+\frac{1}{r^2}\frac{\partial^2 u_\theta}{\partial \theta^2}+\frac{2}{r^2}\frac{\partial u_r}{\partial \theta}\right\}+f_\theta$$

In this chapter, we assume that

- The working fluid has constant density ρ and kinematic viscosity ν (or dynamic viscosity μ) and is Newtonian.
- The flow is laminar and incompressible.
- There is no body force: $\mathbf{f}=0$ or $f_i=0$.

3.1 Plane Couette Flow

The Couette flow is a drag-induced flow either between parallel flat plates (Fig. 3.1) or between concentric rotating cylinders. The drag-induced flow is thus distinguished from the pressure-induced flow such as the Poiseuille flow.

○ Problem definition

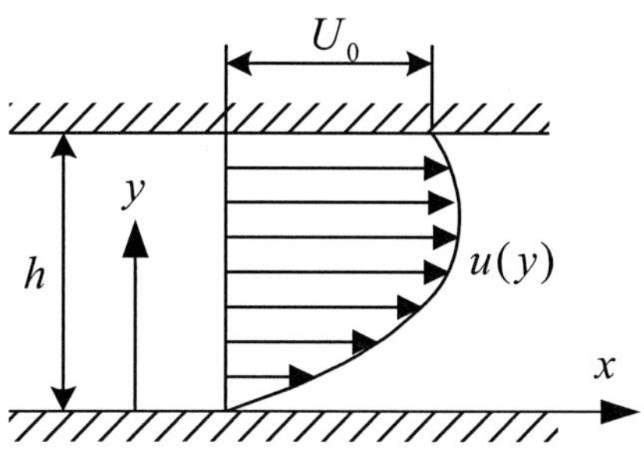

Fig. 3.1 Planar Couette flow within a pair of parallel flat plates.

- Consider a steady laminar flow between two parallel infinite flat plates with a gap h as shown in Fig. 3.1.
- Consider two cases:
  - (i) Both upper and lower plates are stationary.
  - (ii) The lower plate is stationary while the upper plate moves at a constant speed U_0 in the x-direction.
- There is a constant pressure gradient $\partial p/\partial x$ in the x-direction.

○ Deductions

- The flow is steady and invariant in the x- and z-directions:

$$\frac{\partial \mathrm{u}}{\partial t}=0, \quad \frac{\partial \mathrm{u}}{\partial x}=\frac{\partial \mathrm{u}}{\partial z}=0$$

That is, the flow is fully developed in the x-direction.

- There is no flow in the z-direction:

$$w=0, \quad \frac{\partial p}{\partial z}=0$$

○ Governing equations

- Conservation of mass:

From (3.3),

$$0+\frac{\partial v}{\partial y}+0=0 \quad \Rightarrow \quad \frac{\partial v}{\partial y}=0 \quad \Rightarrow v=\text{constant}$$

Since $v=0$ at the boundaries, $v=0$ everywhere.

- x-momentum equation:

  From (3.4a),

$$0+(0+0+0)=-\frac{1}{\rho}\frac{\partial p}{\partial x}+\nu\left(0+\frac{\partial^2 u}{\partial y^2}+0\right)+0 \quad \Rightarrow \frac{1}{\rho}\frac{\partial p}{\partial x}=\nu\frac{\partial^2 u}{\partial y^2}$$

- y-momentum equation:

  From (3.4b),

$$0+(0+0+0)=-\frac{1}{\rho}\frac{\partial p}{\partial y}+\nu(0+0+0)+0 \quad \Rightarrow \frac{\partial p}{\partial y}=0$$

- z-momentum equation:

  From (3.4c),

$$0+(0+0+0)=-\frac{1}{\rho}\frac{\partial p}{\partial z}+\nu(0+0+0)+0 \quad \Rightarrow \frac{\partial p}{\partial z}=0$$

- Pressure:

  Since $\partial p/\partial y=\partial p/\partial z=0$, p is a function of x only: $\partial p/\partial x \rightarrow dp/dx$.

- Simplified equations:

  Since $\partial u/\partial x=\partial u/\partial z=0$, u is a function of y only: $\partial^2 u/\partial y^2 \rightarrow d^2 u/dy^2$. Thus, the exact equations for the plane Couette flow are

$$\nu\frac{d^2 u}{dy^2}=\frac{1}{\rho}\frac{dp}{dx} \tag{3.7}$$

○ Boundary conditions:

We have two different boundary conditions for $u(y)$:

(i) Both upper and lower plates are stationary:

$$\begin{cases} u(0)=0 \\ u(h)=0 \end{cases}$$

(ii) The lower plate is stationary while the upper plate moves at a constant speed U_0 in the x-direction:

$$\begin{cases} u(0)=0 \\ u(h)=U_0 \end{cases}$$

○ Solutions

The simplified equation (3.7) becomes

$$\underbrace{\underbrace{\frac{d^2 u}{dy^2}}_{\text{function of } y \text{ only}} = \underbrace{\frac{1}{\rho\nu}\frac{dp}{dx}}_{\text{function of } x \text{ only}}}_{\text{must be a constant}}$$

Therefore,

$$\frac{d^2u}{dy^2} = \frac{1}{\mu}\frac{dp}{dx} = \text{constant} \quad (\because \ \mu = \rho\nu)$$

Integrating:

$$u = \frac{1}{2\mu}\left(\frac{dp}{dx}\right)y^2 + C_1 y + C_2$$

Here, C_1 and C_2 are to be determined from boundary conditions.

(i) <u>When both upper and lower plates are stationary</u>

$$u = \frac{1}{2\mu}\left(\frac{dp}{dx}\right)y(y-h) \tag{3.8}$$

The maximum velocity $U_{\max}$ can be obtained by setting $du/dy = 0$:

$$U_{\max} = \frac{h^2}{8\mu}\left(-\frac{dp}{dx}\right) \quad \text{at } y = \frac{h}{2}$$

Therefore, we see that $-dp/dx$ must be positive for $U_{\max} > 0$, meaning that the pressure decreases along the x-direction. The volume flow rate per unit width Q is

$$Q = \int_0^h u\,dy = \frac{1}{2\mu}\left(\frac{dp}{dx}\right)\int_0^h (y^2 - hy)\,dy = \frac{2}{3}U_{\max}h$$

(ii) <u>When only the upper plate moves with a constant velocity U_0</u>

$$u = \frac{1}{2\mu}\left(\frac{dp}{dx}\right)y(y-h) + \frac{U_0}{h}y \tag{3.9}$$

The maximum velocity occurs at $y = y_c$, where y_c must satisfy

$$\left(\frac{du}{dy}\right)_{y=y_c} = 0 \ \text{ or } \ y_c = \frac{h}{2} + \frac{2\mu U_0}{h}\left(-\frac{dp}{dx}\right)^{-1}$$

The volume flow rate per unit width Q is

$$Q = \int_0^h u\,dy = \frac{U_0 h}{2} + \frac{h^3}{12\mu}\left(-\frac{dp}{dx}\right)$$

3.2 Poiseuille Flow

The Poiseuille flow is pressure-induced flow (channel flow) in a long duct, usually pipe. The flow is distinguished from the drag-induced flow such as the Couette flow.

○ Problem definition

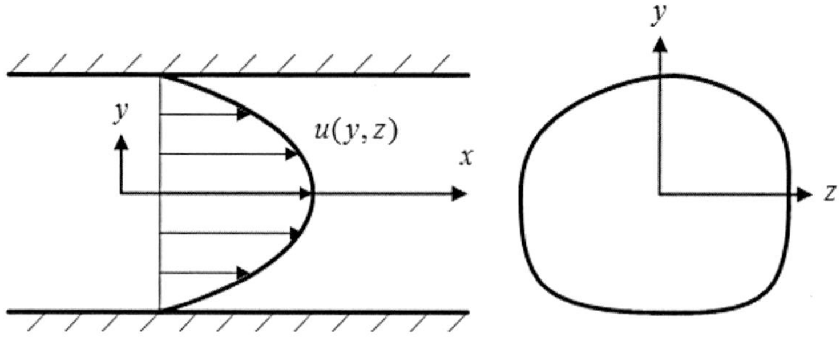

Fig. 3.2 Poiseuille flow within a duct with uniform cross-section.

- Consider a steady laminar flow in an infinitely long pipe (x-direction) with an arbitrary cross section (y-z plane) as shown in Fig. 3.2.
- There is a constant pressure gradient $\partial p/\partial x$ in the x-direction.

○ Deductions:

- The flow is steady:

$$\frac{\partial u}{\partial t}=0$$

- There is no flow in the y-z plane:

$$v=w=0$$

○ Governing equations

- Conservation of mass:

From Eq. (3.3),

$$\frac{\partial u}{\partial x}+0+0=0 \;\Rightarrow\; \frac{\partial u}{\partial x}=0$$

That is, the flow is fully developed in the x-direction. Again, it indicates that u is a function of y and z only: $u=u(y,z)$.

- Conservation of momentum:

From Eq. (3.4),

x-direction:

$$0+(0+0+0)=-\frac{1}{\rho}\frac{\partial p}{\partial x}+\nu\left(0+\frac{\partial^2 u}{\partial y^2}+\frac{\partial^2 u}{\partial z^2}\right)+0$$

$$\Rightarrow \frac{1}{\rho}\frac{\partial p}{\partial x}=\nu\left(\frac{\partial^2 u}{\partial y^2}+\frac{\partial^2 u}{\partial z^2}\right)$$

y-direction:

$$0+(0+0+0)=-\frac{1}{\rho}\frac{\partial p}{\partial y}+\nu(0+0+0)+0 \quad \Rightarrow \frac{\partial p}{\partial y}=0$$

z-direction:

$$0+(0+0+0)=-\frac{1}{\rho}\frac{\partial p}{\partial z}+\nu(0+0+0)+0 \quad \Rightarrow \frac{\partial p}{\partial z}=0$$

- Pressure:

  Since $\partial p/\partial y=\partial p/\partial z=0$, p is a function of x only: $\partial p/\partial x \rightarrow dp/dx$.

- Simplified equations:

  Since u is a function of y and z only, the exact equation for the Poiseulle flow is

$$\frac{\partial^2 u}{\partial y^2}+\frac{\partial^2 u}{\partial z^2}=\frac{1}{\mu}\frac{dp}{dx}=\text{constant} \tag{3.10}$$

  This equation is of Poisson type.

○ Boundary conditions

- The pipe wall is stationary: $u=0$ on the boundary

○ Solutions

There is no general solution for arbitrary cross sections.

(i) For a circular section with a radius a

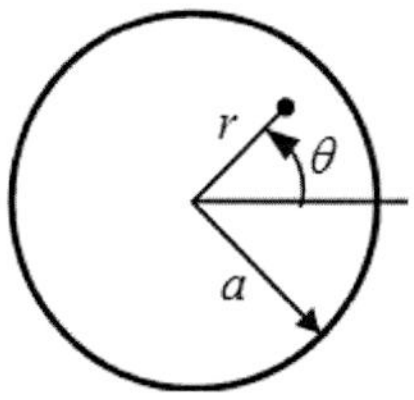

Fig. 3.3 Polar coordinates for the analysis of the Poiseuille flow within a cylindrical duct.

Here, it is more convenient to use the (x,r,θ) coordinate system instead of the (x,y,z) system: $y=r\cos\theta$ and $z=r\sin\theta$, see Fig. 3.3. Since there is no flow on the y-z (or r-θ) plane, $u_r=u_\theta=0$. Because of the axial symmetry, u_x is a function of r only. Therefore, Eq. (3.10) becomes

$$\frac{1}{r}\frac{d}{dr}\left(r\frac{du_x}{dr}\right)=\frac{1}{\mu}\frac{dp}{dx}=\text{constant} \tag{3.11}$$

subjected to the boundary conditions

$$\begin{cases} u_x(0) = \text{finite} \\ u_x(a) = 0 \end{cases}$$

Note that the equation (3.11) can also be directly derived from Eq. (3.6a).

Integrating (3.11) with respect to r:

$$u_x = \frac{1}{4\mu}\left(\frac{dp}{dx}\right)r^2 + \frac{C_1}{\mu}\ln r + C_2$$

Applying the boundary conditions:

$$C_1 = 0 \text{ and } C_2 = -\frac{1}{4\mu}\left(\frac{dp}{dx}\right)a^2$$

Finally we have

$$u_x(r) = \frac{1}{4\mu}\left(-\frac{dp}{dx}\right)(a^2 - r^2) \tag{3.12}$$

Now, we convert this into the Cartesian coordinate system using $r^2 = y^2 + z^2$. Therefore,

$$u(y,z) = \frac{1}{4\mu}\left(-\frac{dp}{dx}\right)a^2\left(1 - \frac{y^2+z^2}{a^2}\right) \tag{3.13}$$

Homework

Using the velocity profile (3.12), obtain the followings:

1) Shear stress distribution, $\tau_{rx}(r)$
2) Shear stress distribution at wall, τ_w
3) Volumetric flow rate, Q
4) Maximum velocity, $U_{\max}$

(ii) For an elliptic cross-section with the major-axis and minor-axis radii a and b

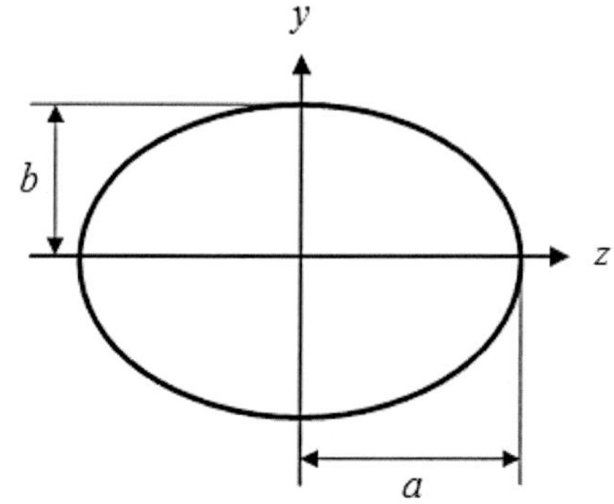

Fig. 3.4 Coordinates for the treatment of the elliptic cross-section.

We can assume

$$u(y,z)=\frac{1}{4\mu}\left(-\frac{dp}{dx}\right)C_1\left(1-\frac{y^2}{a^2}-\frac{z^2}{b^2}\right), \tag{3.14}$$

where C_1 is a constant determined as follows:

$$C_1=\frac{2a^2b^2}{a^2+b^2}$$

Homework

Show that the solution (3.14) satisfies the governing equation (3.10) and derive the formula for C_1. Also show that the solution satisfies the boundary conditions.

3.3 Flow between Rotating Concentric Cylinders

The flow between rotating concentric cylinders is a kind of Couette flow because it is wall-driven flow.

○ Problem definition

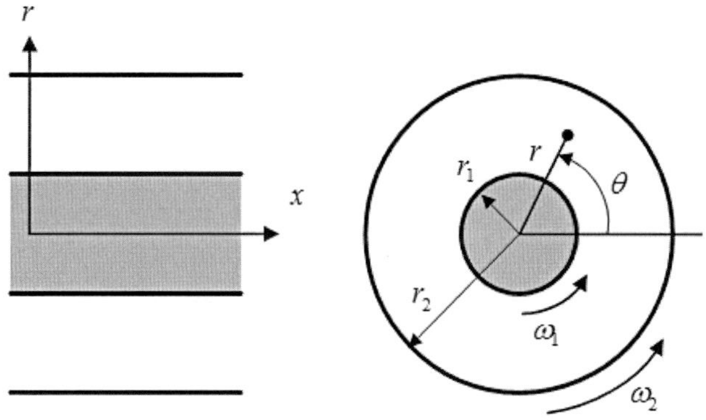

Fig. 3.5 Rotating flow in the annulus between concentric cylinders.

- Consider a steady laminar flow between two infinitely-long concentric cylinders, whose inner and outer surfaces have radii of r_1 and r_2, respectively (Fig. 3.5).
- The inner and outer cylinders are rotating at constant angular velocities ω_1 and ω_2, respectively. The cylinders rotate slowly enough that the creeping flow assumption is valid.

○ Deductions:

- The flow is infinite in the x- and θ-direction and steady:

$$\frac{\partial u}{\partial x}=\frac{\partial u}{\partial \theta}=0\ ;\ \frac{\partial p}{\partial \theta}=0\ ;\ \frac{\partial u}{\partial t}=0$$

- There is no flow in the x- and r-direction:

$u_x = u_r = 0;\ u_\theta = u_\theta(r)$ only

○ Governing equations

- Conservation of mass:

From (3.5),

$0+0+0=0$

The assumptions automatically satisfy the conservation of mass.

- x-momentum equation:

From (3.6a),

$$0+(0+0+0)=-\frac{1}{\rho}\frac{\partial p}{\partial x}+\nu\{0+0+0\}+0 \quad \Rightarrow \frac{\partial p}{\partial x}=0$$

- r-momentum equation:

From (3.6b),

$$0+\left(0+0+0-\frac{u_\theta^2}{r}\right)=-\frac{1}{\rho}\frac{\partial p}{\partial r}+\nu\{0+0+0+0\}+0$$

$$\Rightarrow -\frac{u_\theta^2}{r}=-\frac{1}{\rho}\frac{\partial p}{\partial r}$$

- θ-momentum equation:

From (3.6c),

$$0+(0+0+0+0)=0+\nu\left\{0+\frac{\partial}{\partial r}\left[\frac{1}{r}\frac{\partial(ru_\theta)}{\partial r}\right]+0+0\right\}+0$$

$$\Rightarrow \frac{\partial}{\partial r}\left[\frac{1}{r}\frac{\partial(ru_\theta)}{\partial r}\right]=0$$

- Pressure:

Since $\partial p/\partial x=\partial p/\partial\theta=0$, p is a function of r only: $\partial p/\partial r \rightarrow dp/dr$.

- Simplified equations:

To summarize, the exact equations become

$$\frac{u_\theta^2}{r}=\frac{1}{\rho}\frac{dp}{dr}= \text{ function of } r \text{ only} \tag{3.15}$$

$$\frac{d}{dr}\left[\frac{1}{r}\frac{d(ru_\theta)}{dr}\right]=0 \tag{3.16}$$

○ Boundary conditions

- The inner and outer cylinders rotate at constant angular velocities ω_1 and ω_2, respectively.

$$\begin{cases} u_\theta(r_1) = r_1\omega_1 \\ u_\theta(r_2) = r_2\omega_2 \end{cases}$$

○ Solutions

Integrating (3.16):

$$u_\theta = C_1 r + \frac{C_2}{r} \tag{3.17}$$

Here, C_1 and C_2 are determined from the boundary conditions as follows:

$$C_1 = \frac{\omega_2 r_2^2 - \omega_1 r_1^2}{r_2^2 - r_1^2}$$

$$C_2 = \frac{-r_1^2 r_2^2(\omega_2 - \omega_1)}{r_2^2 - r_1^2}$$

Homework

Answer to the following questions:

1) Obtain the pressure distribution $p(r)$ using Eq. (3.15).

2) Obtain the volume flow rate Q at a radial section.

3) Obtain the shear stresses τ_{w1} and τ_{w2} at the inner and outer surfaces.

4) Obtain the condition (relation between ω_1 and ω_2) for $Q=0$.

3.4 Stokes First Problem

The Stokes first problem concerns the transient flow near an impulsively started flat plate with an infinite length.

○ Problem definition

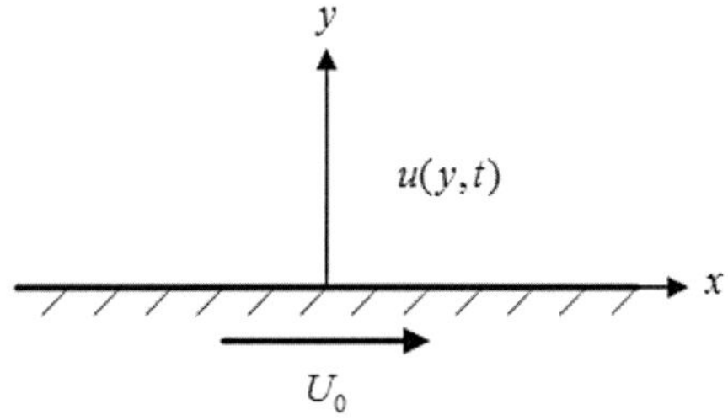

Fig. 3.6 Stokes first problem: parallel flow above a moving flat plate.

- Consider a laminar flow over a flat plate with an infinite length, where the whole fluid is initially at rest before the plate suddenly starts to move (Fig. 3.6).

○ Deductions

- The flow is invariant in the x- and z-direction:

$$\frac{\partial u}{\partial x}=\frac{\partial u}{\partial z}=0\ ;\ u=u(y,t)\ \text{only}$$

- Since the whole fluid will need time to adjust to the plate's motion, the flow is transient (unsteady):

$$\frac{\partial u}{\partial t}\neq 0$$

- There is no flow in the z-direction:

$$w=0$$

- There is no external pressure gradient in the x-direction:

$$\frac{dp}{dx}=0$$

○ Governing equations

- Conservation of mass:

From (3.3),

$$0+\frac{\partial v}{\partial y}+0=0\quad\Rightarrow v=v(t)\ \text{only}$$

Since $v(0,t)=0$, however, $v=0$ everywhere.

- Conservation of momentum:

From (3.4),

y-direction:

$$0+(0+0+0)=-\frac{1}{\rho}\frac{\partial p}{\partial y}+\nu(0+0+0)+0\qquad\Rightarrow\frac{\partial p}{\partial y}=0$$

z-direction:

$$0+(0+0+0)=-\frac{1}{\rho}\frac{\partial p}{\partial z}+\nu(0+0+0)+0\qquad\Rightarrow\frac{\partial p}{\partial z}=0$$

x-direction:

$$\frac{\partial u}{\partial t}+(0+0+0)=-\frac{1}{\rho}\frac{\partial p}{\partial x}+\nu\left(0+\frac{\partial^2 u}{\partial y^2}+0\right)+0$$

$$\Rightarrow\frac{\partial u}{\partial t}=-\frac{1}{\rho}\frac{\partial p}{\partial x}+\nu\frac{\partial^2 u}{\partial y^2}$$

- Pressure:

Since $\partial p/\partial y=\partial p/\partial z=0$, p is a function of x only: $\partial p/\partial x\rightarrow dp/dx$. Since no external pressure gradient is assumed, however, $dp/dx=0$.

- Simplified equations:

Thus, we must solve

$$\frac{\partial u}{\partial t}=\nu\frac{\partial^2 u}{\partial y^2} \tag{3.18}$$

This is the classical diffusion equation with the diffusion coefficient $\nu=[L^2/T]$. Thus, ν represents diffusion of momentum.

○ Boundary and initial conditions

- Fluid moves impulsively at $y=0$:

$$u(0,t)=\begin{cases}U_0 & t>0\\ 0 & t\le 0\end{cases}$$

- Fluid is at rest as $y\to\infty$:

$u(y,t)\to 0$ as $y\to\infty$

- Fluid in the half-plane $y>0$ is initially at rest:

$u(y,0)=0$

○ Solutions: **similarity solution method**

Since the plate is infinite and the fluid domain is semi-infinite, there is no geometric length scale imposed on the solution. Therefore, we can expect a similarity solution. Let

$$\eta=\frac{\alpha y}{t^n} \tag{3.19}$$

where α and n are unknown yet but to be determined later. Further, we assume

$u(y,t)=U_0 f(\eta)$

Then

$$\frac{\partial u}{\partial t}=U_0\frac{\partial f}{\partial t}=U_0\frac{df}{d\eta}\frac{\partial\eta}{\partial t}=U_0 f'(-\alpha n)\frac{y}{t^{n+1}}=U_0 f'\left(-\frac{n}{t}\right)\eta$$

$$\frac{\partial u}{\partial y}=U_0\frac{\partial f}{\partial y}=U_0\frac{df}{d\eta}\frac{\partial\eta}{\partial y}=U_0\left(\frac{\alpha}{t^n}\right)f'$$

$$\frac{\partial^2 u}{\partial y^2}=\frac{\partial}{\partial y}\left(\frac{\partial u}{\partial y}\right)=\frac{\partial}{\partial\eta}\left(\frac{\partial u}{\partial y}\right)\frac{\partial\eta}{\partial y}=U_0\left(\frac{\alpha}{t^n}\right)f''\left(\frac{\alpha}{t^n}\right)=U_0\left(\frac{\alpha}{t^n}\right)^2 f''$$

Substituting all these into the governing equation (3.18), we obtain

$$-\frac{n}{t}U_0\eta f'=\nu U_0\left(\frac{\alpha}{t^n}\right)^2 f'' \quad\text{or}\quad f''+\left(\frac{n}{\nu\alpha^2}\right)t^{2n-1}\eta f'=0$$

In order to have a similarity solution, t^{2n-1} must be constant or $n=1/2$. Further, we choose

$$\left(\frac{n}{\nu\alpha^2}\right)=2 \quad \text{or} \quad \alpha=\frac{1}{2\sqrt{\nu}}$$

so that the governing equation simply becomes

$$f''+2\eta f'=0 \tag{3.20}$$

where the similarity variable is now $\eta=y/2\sqrt{\nu t}$. The boundary conditions are

$$f(0)=1$$

$$f(\eta)\rightarrow 0 \quad \text{as} \quad \eta\rightarrow\infty$$

The solution can be obtained by integration directly

$$\frac{f''}{f'}=-2\eta$$

$$\Rightarrow \ln f'=-\eta^2+C_1'$$

$$\Rightarrow f'=C_1 e^{-\eta^2}$$

$$\Rightarrow f=C_1\int_0^{\eta} e^{-\zeta^2}d\zeta+C_2$$

Applying the boundary condition at $\eta=0$, we get $C_2=1$. Next, apply the boundary condition as $\eta\rightarrow\infty$

$$C_1\int_0^{\infty} e^{-\eta^2}d\eta+1=0$$

But

$$\int_0^{\infty} e^{-\eta^2}d\eta=\frac{\sqrt{\pi}}{2}$$

So,

$$C_1=-\frac{2}{\sqrt{\pi}}$$

Therefore, we have

$$f=1-\underbrace{\frac{2}{\sqrt{\pi}}\int_0^{\eta} e^{-\zeta^2}d\zeta}_{\text{erf}(\eta);\ \text{error function}}=1-\text{erf}(\eta)$$

$$\Rightarrow u=U_0\left[1-\text{erf}(\eta)\right]=U_0\left[1-\text{erf}\left(\frac{y}{2\sqrt{\nu t}}\right)\right] \tag{3.21}$$

Figure 3.7 shows the similarity solution of the velocity profile $u(\eta)$ plotted on the space of the similarity variable η (left) as well as the time evolution of the velocity profile plotted on the physical space y (right) for the Stokes first problem.

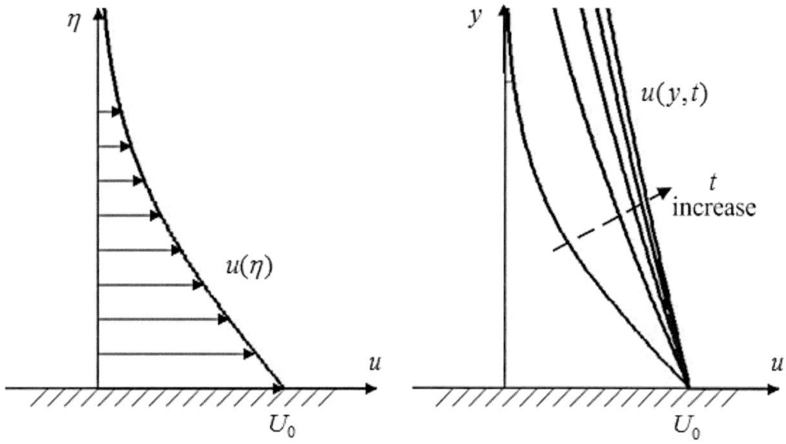

Fig. 3.7 Similarity solution of the velocity profile (left), and the transient velocity profile plotted on the physical space (right).

○ Implications

We can estimate the depth of fluid in time, that is affected by the movement of the plate. Let δ be the y-coordinate at which u is 1% of the wall velocity U_0 and let η_e be the corresponding η. Then

$$0.01\, U_0 = U_0\left[1 - \mathrm{erf}(\eta_e)\right] \quad \text{or} \quad \mathrm{erf}(\eta_e) = 0.99$$

From the mathematical table, we see

$$\eta_e \cong 1.822$$

Also, from $\eta_e = \delta/2\sqrt{\nu t}$, we get $\delta = \left(2\sqrt{\nu}\,\eta_e\right)\sqrt{t}$. This implies that δ increases like $\sqrt{t}$. This result also shows that as ν becomes smaller (less viscous), δ becomes thinner at the same time t.

3.5 Stokes Second Problem

The Stokes second problem concerns unsteady flow near a flat plate with an infinite length which oscillates in time. Except for the boundary condition at the plane wall, the Stokes second problem is exactly identical to the Stokes first one.

○ Problem definition

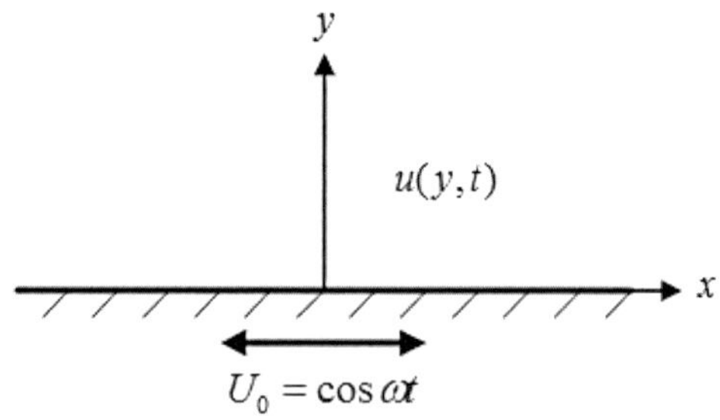

Fig. 3.8 Stokes second problem: parallel flow above an oscillating flat plate

- Consider a laminar flow over a flat plate with infinite length, where the plate oscillates with an angular frequency ω (Fig. 3.8).

○ Governing equations

- Simplified equations:

We must solve the same governing equation as that for the Stokes first problem:

$$\frac{\partial u}{\partial t} = \nu \frac{\partial^2 u}{\partial y^2} \tag{3.18}$$

○ Boundary conditions

- Fluid oscillates in a sinusoidal mode in time at $y=0$:

$$u(0,t) = U_0 \cos \omega t$$

- Fluid is at rest as $y \rightarrow \infty$:

$$u(y,t) \rightarrow 0 \text{ as } y \rightarrow \infty$$

○ Solutions:

To solve the linear governing equation (3.18), we assume

$$u(y,t) = \text{Real}\left[f(y)e^{i\omega t}\right] = f(y)\cos \omega t$$

Since the equation is linear, we can just throw out the symbol "Real" and at the final stage we just take only the real part.

$$\frac{\partial u}{\partial t} = i\omega f e^{i\omega t}$$

$$\frac{\partial^2 u}{\partial y^2} = f'' e^{i\omega t}$$

Therefore, the governing equation becomes

$$f'' - i\frac{\omega}{\nu}f = 0$$

Here, the characteristic equation is

$$\lambda^2 - i\frac{\omega}{\nu} = 0 \quad \text{or}$$

$$\lambda = \pm\sqrt{\frac{\omega}{\nu}}\sqrt{i} = \pm\sqrt{\frac{\omega}{\nu}}\frac{1+i}{\sqrt{2}} = \pm\sqrt{\frac{\omega}{2\nu}}(1+i)$$

Then we arrive at

$$f(y) = A\exp\left[-(1+i)\sqrt{\frac{\omega}{2\nu}}y\right] + B\exp\left[(1+i)\sqrt{\frac{\omega}{2\nu}}y\right]$$

To satisfy the boundary condition at infinity, we set $B=0$. Then,

$$f(y) = A\exp\left(-\sqrt{\frac{\omega}{2\nu}}y\right)\exp\left(-i\sqrt{\frac{\omega}{2\nu}}y\right)$$

Therefore,

$$u(y,t) = A\exp\left(-\sqrt{\frac{\omega}{2\nu}}y\right)\exp\left[i\left(\omega t - \sqrt{\frac{\omega}{2\nu}}y\right)\right]$$

Finally, we take a real part;

$$u(y,t) = A\exp\left(-\sqrt{\frac{\omega}{2\nu}}y\right)\cos\left(\omega t - \sqrt{\frac{\omega}{2\nu}}y\right)$$

To satisfy the boundary condition at $y=0$, we must take $A = U_0$. Therefore,

$$u = \underbrace{U_0\exp\left(-\sqrt{\frac{\omega}{2\nu}}y\right)}_{\text{amplitude}}\cos\left(\omega t - \sqrt{\frac{\omega}{2\nu}}y\right) \tag{3.22}$$

Figure 3.8 shows the time evolution of the velocity profile (3.22) for the Stokes second problem.

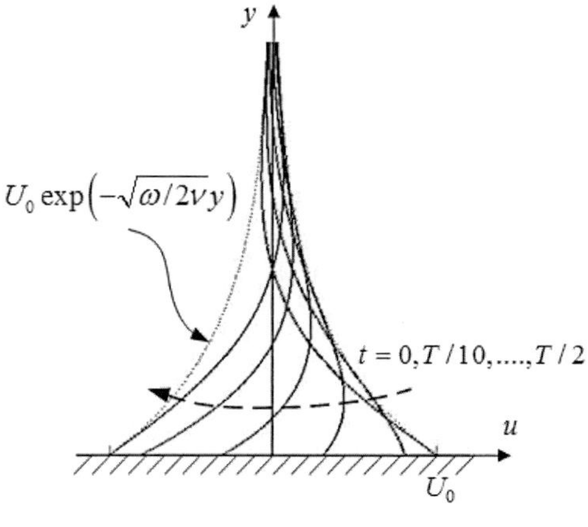

Fig. 3.9 Oscillatory velocity profile plotted on the physical space, y.

○ Implications:

The wave length λ for a fixed time is $\lambda = 2\pi\sqrt{2\nu/\omega}$. Let δ be the y-value at which the amplitude of u becomes 1% of U_0:

$$0.01 = \exp\left(-\sqrt{\frac{\omega}{2\nu}}\,\delta\right) \quad \text{or}$$

$$\delta = -(\ln 0.01)\sqrt{\frac{2\nu}{\omega}} = 4.605\sqrt{\frac{2\nu}{\omega}}$$

Thus, we can see that the thickness of the fluid layer, δ, that feels as much as 1% of the plate motion, increases with ν like $\sqrt{\nu}$ and decreases with ω like $1/\sqrt{\omega}$. This result also shows that as the fluid becomes less viscous, the layer becomes thinner. In addition, as the frequency becomes higher, the layer also becomes thinner.

3.6 Pulsating Flow between Parallel Surfaces

○ Problem definition

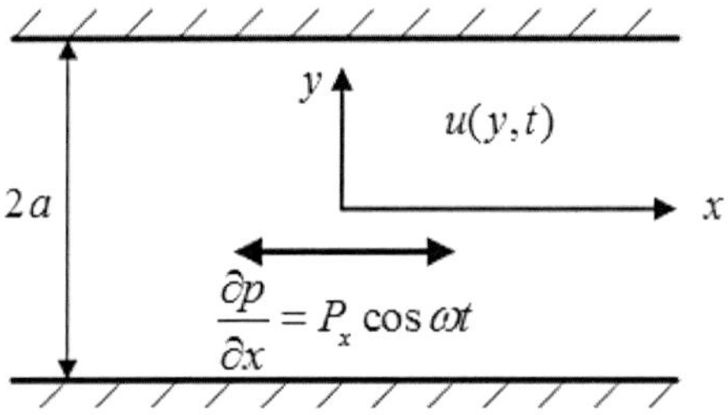

Fig. 3.10 Pulsating flow driven by the fluid pressure force showing harmonic variation in time between parallel plates.

- Consider a laminar flow bounded by two parallel planes with infinite area and a gap $2a$, where an external pressure gradient in the streamwise (x) direction oscillates in time with a frequency ω (Fig. 3.10).

○ Deductions

- The flow is invariant in the x- and z-direction:

$$\frac{\partial u}{\partial x} = \frac{\partial u}{\partial z} = 0 \;;\; u = u(y,t) \text{ only}$$

- Since the fluid pulsates in time, the flow is unsteady:

$$\frac{\partial u}{\partial t} \neq 0$$

- There is no flow in the z-direction:

  $w=0$

- There is an external pressure gradient oscillating in time imposed in the x -direction:

$$\frac{\partial p}{\partial x}=P_x\cos\omega t$$

○ Governing equations

- Conservation of mass:

  From (3.3),

$$0+\frac{\partial v}{\partial y}+0=0 \quad \Rightarrow \quad \frac{\partial v}{\partial y}=0 \quad \Rightarrow v=v(t) \ \text{ only}$$

  Since $v=0$ at $y=\pm a$, $v=0$ always everywhere.

- Conservation of momentum:

  From (3.4),

  x-direction:

$$\frac{\partial u}{\partial t}+(0+0+0)=-\frac{1}{\rho}\frac{\partial p}{\partial x}+\nu\left(0+\frac{\partial^2 u}{\partial y^2}+0\right)+0$$

$$\Rightarrow \quad \frac{\partial u}{\partial t}=-\frac{1}{\rho}\frac{\partial p}{\partial x}+\nu\frac{\partial^2 u}{\partial y^2}$$

  y-direction:

$$0+(0+0+0)=-\frac{1}{\rho}\frac{\partial p}{\partial y}+\nu(0+0+0)+0 \quad \Rightarrow \quad \frac{\partial p}{\partial y}=0$$

  z-direction:

$$0+(0+0+0)=-\frac{1}{\rho}\frac{\partial p}{\partial z}+\nu(0+0+0)+0 \quad \Rightarrow \quad \frac{\partial p}{\partial z}=0$$

- Pressure:

  Since $\partial p/\partial y=\partial p/\partial z=0$, p is a function of x only: $\partial p/\partial x \rightarrow dp/dx$. Since an oscillating pressure gradient is externally imposed, we can write

$$\frac{dp}{dx}=P_x\cos\omega t$$

- Simplified equations:

  Thus, we must solve

$$\frac{\partial u}{\partial t}=-\frac{1}{\rho}P_x\cos\omega t+\nu\frac{\partial^2 u}{\partial y^2} \tag{3.23}$$

○ Boundary conditions:

- The plane surfaces are stationary: $u(\pm a,t)=0$

○ Solutions:

Since the governing equation is linear, we can expect a separable solution for $u(y,t)$. As before, we assume

$$u(y,t) = \text{Real}\left[f(y)e^{i\omega t}\right]$$

Then by substituting this and its derivatives into the governing equation (3.23), we obtain

$$i\omega f e^{i\omega t} = -\frac{1}{\rho}P_x e^{i\omega t} + \nu f'' e^{i\omega t} \qquad \text{or} \quad f'' - \frac{\omega}{\nu} i f = \frac{P_x}{\rho\nu}$$

The particular solution to this ordinary differential equation is

$$f_p = \frac{P_x}{\rho\omega} i$$

The complementary solution is

$$f_c = A\cosh\left[(1+i)\sqrt{\frac{\omega}{2\nu}}\,y\right] + B\sinh\left[(1+i)\sqrt{\frac{\omega}{2\nu}}\,y\right]$$

Then, the general solution is the sum of particular and complementary solutions;

$$f = f_p + f_c$$
$$= \frac{P_x}{\rho\omega} i + A\cosh\left[(1+i)\sqrt{\frac{\omega}{2\nu}}\,y\right] + B\sinh\left[(1+i)\sqrt{\frac{\omega}{2\nu}}\,y\right]$$

For the boundary conditions to be satisfied, we must set

$$A = \frac{-iP_x}{\rho\omega\cosh\left[(1+i)\sqrt{\frac{\omega}{2\nu}}\,a\right]}$$

$$B = 0$$

Thus, the final solution is

$$u(y,t) = \text{Real}\left\{ i\frac{P_x}{\rho\omega}\left[1 - \frac{\cosh\left[(1+i)\sqrt{\frac{\omega}{2\nu}}\,y\right]}{\cosh\left[(1+i)\sqrt{\frac{\omega}{2\nu}}\,a\right]}\right] e^{i\omega t}\right\} \tag{3.24}$$

This expression may be decomposed to yield the real part explicitly. Although the concepts are straightforward, the details are cumbersome.

Homework

Answer to the following questions:

1) Plot the velocity distribution for $t = 0$, $T/8$, $2T/8$, $3T/8$ and $4T/8$.

2) Discuss the nature of the flow.

3.7 Stagnation-Point Flow (Hiemenz Flow)

A stagnation point occurs whenever a flow impinges on a solid object (Fig. 3.11). Flow around the stagnation point is referred to as the stagnation-point flow.

○ Problem definition

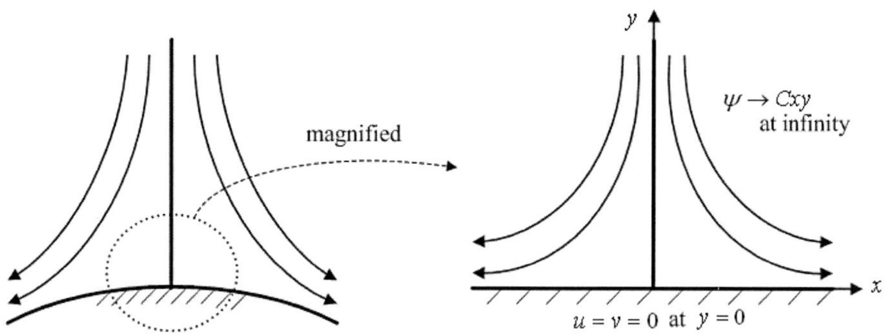

Fig. 3.11 Stagnation point defined in a flow approaching to an obstacle (left) and the stagnation-point (or Hiemenz) flow defined as the asymptote of the flow around the stagnation point (right).

- Consider a steady laminar flow that impinges with a constant velocity U_0 on a flat plane boundary at a right angle.

○ Deductions

- The flow is steady and invariant in the z-direction:

$$\frac{\partial \mathrm{u}}{\partial z}=0\,;\ \frac{\partial \mathrm{u}}{\partial t}=0$$

- There is no flow in the z-direction:

$$w=0$$

- The flow is assumed to be inviscid far from the boundary wall:

Since the wall permits the slip condition, then the flow field is potential:

$$\psi = 2U_0 xy \tag{3.25}$$

$$u=\frac{\partial \psi}{\partial y}=2\,U_0 x$$

$$v=-\frac{\partial \psi}{\partial y}=-2U_0 y$$

Here, ψ is the stream function. From the Bernoulli equation,

$$p=p_o-2\rho U_0^2\left(x^2+y^2\right) \tag{3.26}$$

At the stagnation point $(x=y=0)$, $p=p_0$. Thus, p_0 corresponds to the stagnation pressure.

Homework

In the impinging inviscid flow over a plane boundary, show that the stream function is a form of $\psi=2U_0xy$.

○ Boundary condition

- The bottom wall is stationary:

$$u=v=0 \quad \text{at } y=0$$

○ Governing equations

- Modification of the potential flow:

This potential flow must be modified to satisfy the no-slip condition at the wall. Assume

$$u=2U_0xf_1'(y)$$

Then,

$$\psi=\int u\,dy=2U_0xf_1(y)+C_1(x)$$

$$v=-\frac{\partial\psi}{\partial x}=-2U_0f_1(y)-C_1'(x)$$

Apply the boundary condition for u

$$f_1'(0)=0$$

Apply the boundary condition for v

$$-2U_0f_1(0)-C_1'(x)=0$$

$$\Rightarrow C_1'(x)=-2U_0f_1(0)$$

$$\Rightarrow C_1(x)=-2U_0xf_1(0)$$

Therefore,

$$\psi=2U_0x\left[f_1(y)-f_1(0)\right]$$

Let $f_1(y)-f_1(0)=f(y)$. Then $f_1'(y)=f'(y)$. Finally we have

$$u=2U_0xf'(y)$$

$$v=-2U_0f(y)$$

subjected to the boundary conditions

$$f(0)=f'(0)=0 \quad \text{and} \quad f'(y)\rightarrow 1 \text{ as } y\rightarrow\infty$$

- Conservation of momentum:

We substitute the forms for (u,v) to the Navier-Stokes equations (3.4):

x-direction:

$$0+\left[(2U_0xf')2U_0f'+(-2U_0f)2U_0xf''+0\right]$$
$$=-\frac{1}{\rho}\frac{\partial p}{\partial x}+\nu\left(0+2U_0xf'''+0\right)+0$$

$$\Rightarrow 4U_0^2x(f')^2-4U_0^2xff''=-\frac{1}{\rho}\frac{\partial p}{\partial x}+2\nu U_0xf'''$$

y-direction:

$$0+\left[0+(-2U_0f)(-2U_0f')+0\right]=-\frac{1}{\rho}\frac{\partial p}{\partial y}+\nu\left(0-2U_0f''+0\right)+0$$

$$\Rightarrow 4U_0^2ff'=-\frac{1}{\rho}\frac{\partial p}{\partial y}-2\nu U_0f''$$

z-direction:

$$0+(0+0+0)=-\frac{1}{\rho}\frac{\partial p}{\partial z}+\nu(0+0+0)+0 \quad \Rightarrow \frac{\partial p}{\partial z}=0$$

- Pressure:

  We can integrate the y-momentum equation to get

  $$p=2\rho U_0^2f^2-2\nu\rho U_0f'+g(x)$$

  At infinity, we must recover the potential-flow pressure (3.26);

  $$p=p_0-2\rho U_0^2(x^2+y^2)$$

  On the other hand, at infinity, $f'\rightarrow 1$ and so $f\rightarrow y+\text{constant}$. Comparing the two forms for p, we get $g(x)=-2\rho U_0^2x^2+\text{constant}$. Therefore,

  $$p=-2\rho U_0^2f^2-2\nu\rho U_0f'-2\rho U_0^2x^2+\text{constant}$$

  $$\frac{\partial p}{\partial x}=-4\rho U_0^2x$$

- Simplified equations:

  Substitute the pressure gradient into the x-momentum equation to get

  $$4U_0^2xf'^2-4U_0^2xff''=4U_0^2x+2U_0\nu xf'''$$

  Therefore,

  $$\frac{\nu}{2U_0}f'''+ff''+\left(1-f'^2\right)=0 \qquad (3.27)$$

○ Boundary condition: revisited

$$f(0)=f'(0)=0 \quad \text{and} \quad f'(y)\rightarrow 1 \ \text{as} \ y\rightarrow\infty$$

○ Solutions:

Let

$$\eta = \sqrt{\frac{2U_0}{\nu}}\, y \quad \text{and} \quad \phi(\eta) = \sqrt{\frac{2U_0}{\nu}}\, f(\eta)$$

Then,

$$\frac{df}{dy} = \frac{df}{d\eta}\frac{d\eta}{dy} = \sqrt{\frac{2U_0}{\nu}}\frac{df}{d\eta} = \frac{d\phi}{d\eta} = \phi'$$

$$\frac{d^2 f}{dy^2} = \frac{d}{dy}\left(\frac{df}{dy}\right) = \frac{d}{dy}(\phi') = \frac{d}{dy}(\phi')\frac{d\eta}{dy} = \sqrt{\frac{2U_0}{\nu}}\phi''$$

$$\frac{d^3 f}{dy^3} = \frac{2U_0}{\nu}\phi'''$$

Therefore,

$$\phi''' + \phi\phi'' + \left(1 - \phi'^2\right) = 0 \qquad (3.28)$$

subjected to the boundary conditions

$\phi(0) = \phi'(0) = 0$ and

$\phi'(\eta) \to 1$ as $\eta \to \infty$

We must use a numerical scheme to solve this ordinary differential equation. It is found that

$$\phi''(0) = 1.232588$$

$$\lim_{\eta \to \infty} \phi(\eta) = \eta - 0.647900$$

Figure 3.12 shows the schematic functional forms of solution for the stagnation-point flow: $\phi(\eta)$ and $\phi'(\eta)$.

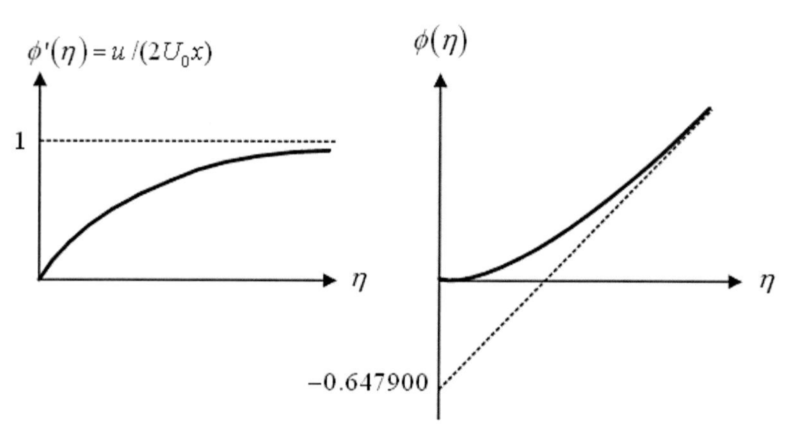

Fig. 3.12 Schematic of the functional forms of similarity solution for the stagnation-point flow: $\phi(\eta)$ and $\phi'(\eta)$.

○ Implications:

Let δ be the y-value at which u reaches 99% of the potential-flow velocity $2U_0x$, that is

$$0.99 = \phi'(\eta_e) = \phi'\left(\sqrt{\frac{2U_0}{\nu}}\,\delta\right)$$

From the numerical result, it is found that

$$\sqrt{\frac{2U_0}{\nu}}\,\delta \simeq 2.4 \quad \text{or} \quad \delta = 2.4\sqrt{\frac{\nu}{2U_0}}$$

This implies that δ increases with ν like $\sqrt{\nu}$. In other words, as the fluid becomes more viscous, the layer becomes thicker. Figure 3.13 shows schematic velocity profiles and boundary thickness for the stagnation-point flow.

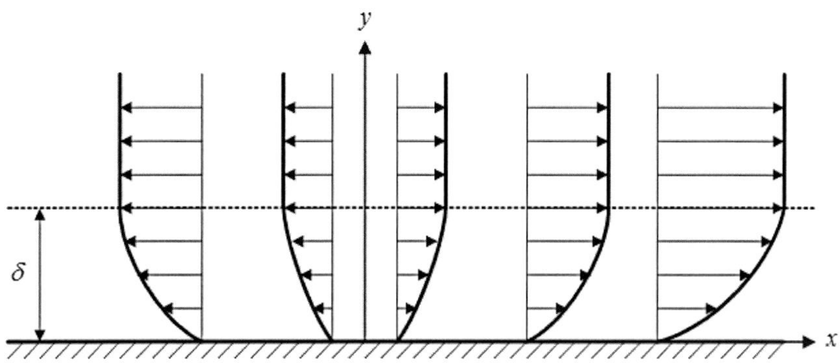

Fig. 3.13 Velocity profiles of the stagnation-point flow at several locations of x.

3.8 Flow in Convergent and Divergent Channels

○ Problem definition

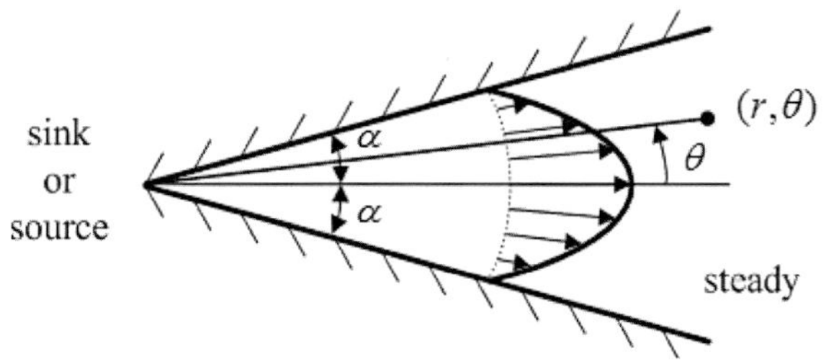

Fig. 3.14 Coordinates and geometry of the convergent (or divergent) channel with a sink (or source). The sketch of the velocity profile shown corresponds to the divergent flow.

- Consider a steady laminar flow in a convergent (divergent) channel with a converging (diverging) angle α (Fig. 3.14).
- Figure 3.14 depicts flow in a divergent channel: $-\alpha \le \theta \le \alpha$ $(0 < \alpha < \pi/2)$. To consider flow in a convergent channel, we have to change the range to $\pi - \alpha \le \theta \le \pi + \alpha$.

○ Deductions

- The flow domain is steady and invariant in the x-direction (normal to the plane):

$$\frac{\partial \mathrm{u}}{\partial x} = 0, \quad \frac{\partial \mathrm{u}}{\partial t} = 0$$

- There is no flow in the x- and θ-direction:

$$u_x = u_\theta = 0, \quad u_r = u_r(r,\theta) \text{ only}$$

○ Governing equations

- Conservation of mass:

From (3.5),

$$0 + \frac{1}{r}\frac{\partial (ru_r)}{\partial r} + 0 = 0 \quad \Rightarrow \quad \frac{1}{r}\frac{\partial (ru_r)}{\partial r} = 0$$

- x-momentum equation:

From (3.6a),

$$0 + (0+0+0) = -\frac{1}{\rho}\frac{\partial p}{\partial x} + \nu\{0+0+0\} + 0 \quad \Rightarrow \quad \frac{\partial p}{\partial x} = 0$$

- r-momentum equation:

From (3.6b),

$$0 + \left(0 + u_r \frac{\partial u_r}{\partial r} + 0 - 0\right) = -\frac{1}{\rho}\frac{\partial p}{\partial r}$$

$$+ \nu\left\{0 + \frac{\partial}{\partial r}\left[\frac{1}{r}\frac{\partial (ru_r)}{\partial r}\right] + \frac{1}{r^2}\frac{\partial^2 u_r}{\partial \theta^2} + 0\right\} + 0$$

$$\Rightarrow u_r \frac{\partial u_r}{\partial r} = -\frac{1}{\rho}\frac{\partial p}{\partial r} + \nu\left\{\frac{\partial}{\partial r}\left[\frac{1}{r}\frac{\partial (ru_r)}{\partial r}\right] + \frac{1}{r^2}\frac{\partial^2 u_r}{\partial \theta^2}\right\}$$

- θ-momentum equation:

From (3.6c),

$$0 + (0+0+0+0) = -\frac{1}{\rho}\frac{1}{r}\frac{\partial p}{\partial \theta} + \nu\left\{0+0+0+\frac{2}{r^2}\frac{\partial u_r}{\partial \theta}\right\} + 0$$

$$\Rightarrow 0 = -\frac{1}{\rho}\frac{1}{r}\frac{\partial p}{\partial \theta} + \nu\left\{\frac{2}{r^2}\frac{\partial u_r}{\partial \theta}\right\}$$

- Simplified equations:

$$\frac{1}{r}\frac{\partial(ru_r)}{\partial r}=0 \tag{3.29}$$

$$u_r\frac{\partial u_r}{\partial r}=-\frac{1}{\rho}\frac{\partial p}{\partial r}+\nu\left\{\frac{\partial}{\partial r}\left[\frac{1}{r}\frac{\partial(ru_r)}{\partial r}\right]+\frac{1}{r^2}\frac{\partial^2 u_r}{\partial\theta^2}\right\} \tag{3.30}$$

$$0=-\frac{1}{\rho}\frac{1}{r}\frac{\partial p}{\partial\theta}+\nu\left\{\frac{2}{r^2}\frac{\partial u_r}{\partial\theta}\right\} \tag{3.31}$$

○ Boundary conditions:

- The channel wall is stationary;

$u_r(r,\theta)=0$ at $\theta=\pm\alpha$

○ Solutions:

In this problem, there is no specific length scale. However, θ itself acts as a similarity variable. Thus, we seek a separable-form solution:

$$u_r=R(r)\Theta(\theta)$$

Substituting this into the continuity equation (3.29), we get

$$\frac{\partial}{\partial r}[rR(r)\Theta(\theta)]=0 \text{ or } R(r)=\frac{\text{constant}}{r}$$

We choose ν for the constant so that $\Theta(\theta)$ becomes dimensionless;

$$u_r=\frac{\nu}{r}\Theta(\theta)$$

Next, we substitute this into the momentum equations (3.30) and (3.31) to get

$$-\frac{\nu^2}{r^2}\Theta^2=-\frac{1}{\rho}\frac{\partial p}{\partial r}+\frac{\nu^2}{r^3}\Theta''$$

$$0=-\frac{1}{\rho r}\frac{\partial p}{\partial\theta}+2\frac{\nu^2}{r^3}\Theta'$$

We eliminates the pressure terms by taking $\partial/\partial\theta$ for the first equation and $r(\partial/\partial r)$ for the second and subtracting one from the other. The result is

$$\Theta'''+4\Theta'+2\Theta\Theta'=0$$

This equation may be immediately integrated once

$$\Theta''+4\Theta+\Theta^2=K$$

subjected to the boundary conditions

$\Theta(\alpha)=\Theta(-\alpha)=0$ (divergent)

$\Theta'(0)=0$

where K is a constant of integration.

K is related to the pressure gradient on the wall. From the r-momentum equation, we see that

$$\frac{\partial p}{\partial r} = \frac{\rho\nu^2}{r^3}\Theta'' \quad \text{at} \quad \theta = \pm\alpha$$

Since $(\Theta'')_w = K$ from the above equation for Θ, this can be written in terms of K as

$$\left(\frac{\partial p}{\partial r}\right)_w = \frac{\rho\nu^2}{r^3}K$$

Figure 3.15 shows schematic velocity profiles for the flow in convergent and divergent channels.

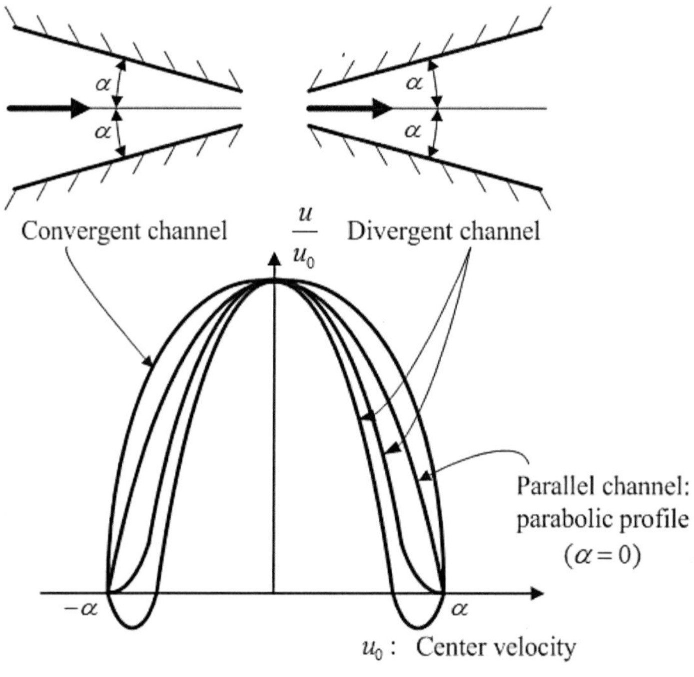

Fig. 3.15 Velocity profiles for various values of α in the convergent (left) and divergent (right) channel.

Chapter 4. Heat Transfer

4.1 Introduction

○ Heat is transferred through three (main) mechanisms

- Conduction: due to the molecular activity
- Convection: conveyed by fluid motion
- Radiation: due to the energy emission from surface

4.2 Conduction

The heat transfer rate (amount of heat energy passing through a given surface per unit time) q_x is obtained by

$$q_x = kA \lim_{\Delta x \to 0} \frac{\Delta T}{\Delta x} = kA\frac{dT}{dx} \quad \text{(Fourier's law)}$$

where k is the thermal conductivity of the material used in the heat passing and A is the area (Fig. 4.1).

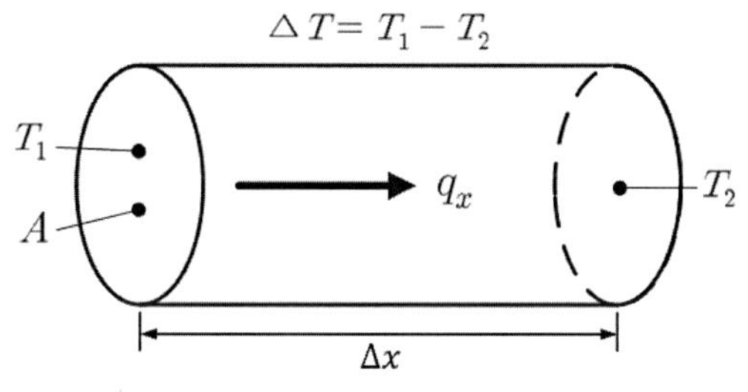

Fig. 4.1 Heat flow along the x-direction in the space where temperature T varies.

We define the heat flux (heat transfer rate per unit area) for the one-diemnsional problem as

$$q'' = k\frac{dT}{dx}$$

In general, for 3-D,

$$\mathrm{q}'' = -k\nabla T \tag{4.1}$$

Note that the heat flux is a vector quantity. We need '–' sign because heat is flowing from hot to cold regions.

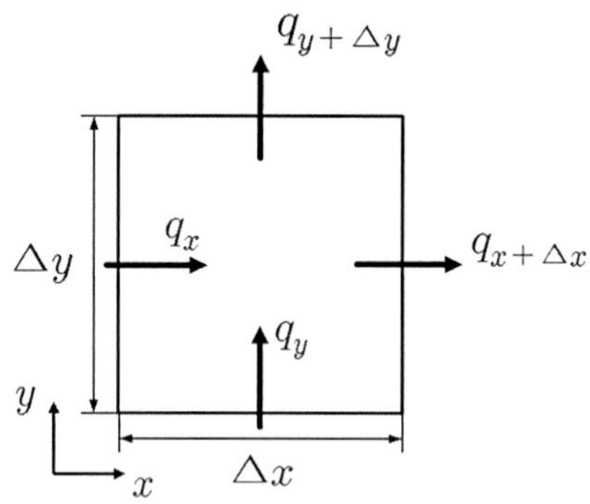

Fig. 4.2 Infinitesimal material element in a 2-D space and heat flux through the surrounding surfaces.

We now derive the heat diffusion equation. For a 2-D case, shown in Fig. 4.2, the heat transfer rate through the left-hand side surface is

$$q_x = -k\Delta y\frac{\partial T}{\partial x}$$

Similarly, $q_{x+\Delta x}$ at the right-hand side surface is ;

$$\begin{aligned} q_{x+\Delta x} &= q_x + \frac{\partial q_x}{\partial x}\Delta x \\ &= -k\Delta y\frac{\partial T}{\partial x} + \frac{\partial}{\partial x}\left(-k\Delta y\frac{\partial T}{\partial x}\right)\Delta x \\ &= -k\Delta y\frac{\partial T}{\partial x} - \Delta y\Delta x\frac{\partial}{\partial x}\left(k\frac{\partial T}{\partial x}\right) \end{aligned}$$

Similarly, on the lower surface;

$$q_y = -k\Delta x\frac{\partial T}{\partial y}$$

On the upper surface;

$$q_{y+\Delta y} = -k\Delta x\frac{\partial T}{\partial y} - \Delta x\Delta y\frac{\partial}{\partial y}\left(k\frac{\partial T}{\partial y}\right)$$

The net amount of heat transfer rate, which is put into the domain, must be accumulated in it;

$$q_x + q_y - q_{x+\Delta x} - q_{y+\Delta y} = \rho c_p\frac{\partial T}{\partial t}\Delta x\Delta y$$

which leads to

$$\frac{\partial}{\partial x}\left(k\frac{\partial T}{\partial x}\right) + \frac{\partial}{\partial y}\left(k\frac{\partial T}{\partial y}\right) = \rho c_p\frac{\partial T}{\partial t} \tag{4.2}$$

Here, ρ is the density of fluid and c_p is the heat capacity in $[\mathrm{J/kgK}]$; the latter can be understood as the heat energy needed to increase the unit temperature of $1\mathrm{kg}$ mass.

Note that Eq. (4.2) is also derived in studying the Laminar Flow (Chapter 2).

For 3-D and constant thermal conductivity k,

$$\frac{\partial T}{\partial t} = \alpha \nabla^2 T \tag{4.3}$$

where $\alpha = k/\rho c_p$ is the thermal diffusivity. When there is heat source in the domain,

$$\frac{\partial T}{\partial t} = \alpha \nabla^2 T + \frac{\alpha}{k} \dot{q} \tag{4.4}$$

where $\dot{q}$ denotes the amount of heat generated within the domain per unit volume and unit time.

4.3 One-Dimensional Steady-State Conduction

For 1-D steady-state conduction through a solid slab without any heat source (Fig. 4.3), we can write for $T = T(x)$ as

$$\frac{d}{dx}\left(k \frac{dT}{dx}\right) = 0$$

We apply the boundary conditions

$$\begin{cases} T = T_{s,1} \text{ at } x = 0 \\ T = T_{s,2} \text{ at } x = L \end{cases}$$

The solution is

$$T(x) = (T_{s,2} - T_{s,1}) \frac{x}{L} + T_{s,1}$$

showing a linear distribution in space. The heat transfer rate is then given as

$$q_x = -kA \frac{dT}{dx} = \frac{kA}{L} (T_{s,1} - T_{s,2})$$

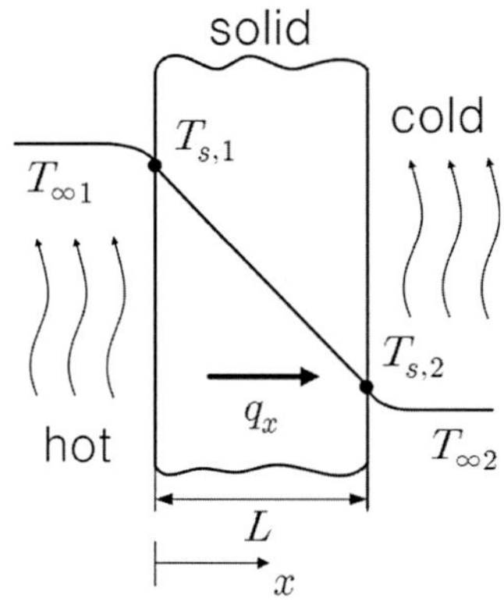

Fig. 4.3 1-D heat transfer through a slab of thickness L surrounded by hot and cold fluids.

This can be understood as follows. The heat transfer rate (corresponding to current in electricity) is proportional to the temperature difference (corresponding to voltage difference) and the proportional constant is kA/L; therefore, the thermal resistance is L/kA ($1/k \rightarrow$ resistivity). The equivalent electric circuit is shown in Fig. 4.4.

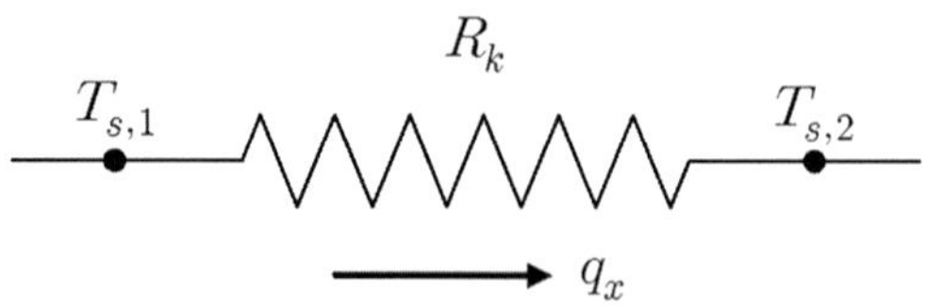

Fig. 4.4 Equivalent electric circuit for the heat transfer in the slab.

When a hot fluid is in contact with the slab at $x=0$, heat transfer occurs from the fluid to the solid (Fig. 4.5). For now, we simply write the heat transfer rate as

$$q_x = h_1 A(T_{\infty 1} - T_{s,1}) \tag{4.5}$$

where h_1 is the convective heat transfer coefficient (see section 4.6 for details). Then, the thermal resistance is $1/(h_1 A)$. Similarly, the fluid maintained at a temperature $T_{\infty 2}$ lower than the solid slab receives heat from the solid in the rate

$$q_x = h_2 A(T_{s,2} - T_{\infty 2}) \tag{4.6}$$

which indicates that thermal resistance is $1/(h_2 A)$.

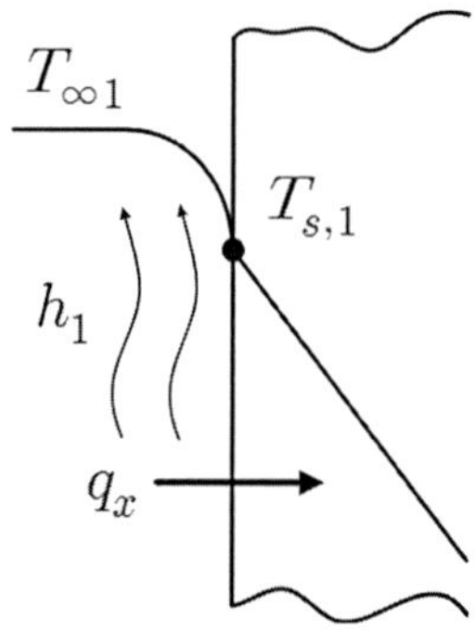

Fig. 4.5 Heat transfer at the interface from the fluid to the solid.

○ Equivalent electric circuit (thermal circuit)

Figure 4.6 shows the equivalent electric circuit of the 1-D heat transfer problem discussed so far.

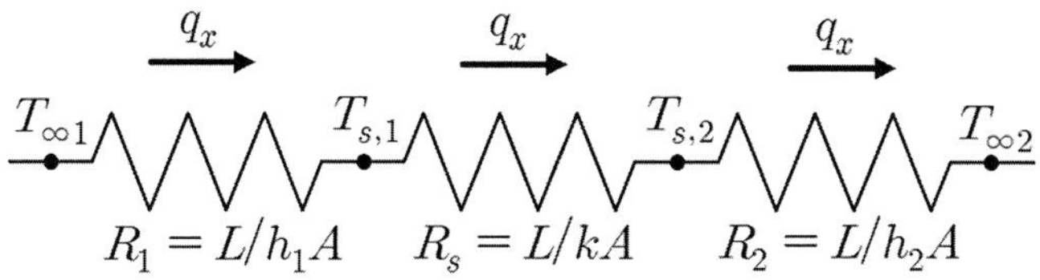

Fig. 4.6 Equivalent electric circuit for the heat transfer in the full domain of Fig. 4.3.

Note that q_x is the same at every resistance. Total resistance is

$$R_{tot} = R_1 + R_s + R_2 = \frac{1}{h_1 A} + \frac{L}{kA} + \frac{1}{h_2 A} \tag{4.7}$$

The heat transfer rate is then

$$q_x = \frac{T_{\infty 1} - T_{\infty 2}}{R_{tot}} = \frac{A(T_{\infty 1} - T_{\infty 2})}{1/h_1 + L/k + 1/h_2} \tag{4.8}$$

The temperatures at the junctions can be found from

$$q_x = h_1 A(T_{\infty 1} - T_{s,1})$$
$$q_x = h_2 A(T_{s,2} - T_{\infty 2})$$

○ Contact resistance

In many cases, there is a temperature drop across an interface between materials (Fig. 4.7), which is attributed to thermal contact resistance R_{tc}. So, the heat transfer rate across the interface can be written as

$$q_x = \frac{A}{R_{tc}}(T_A - T_B) \tag{4.9}$$

The existence of the temperature drop is due to the surface roughness effect.

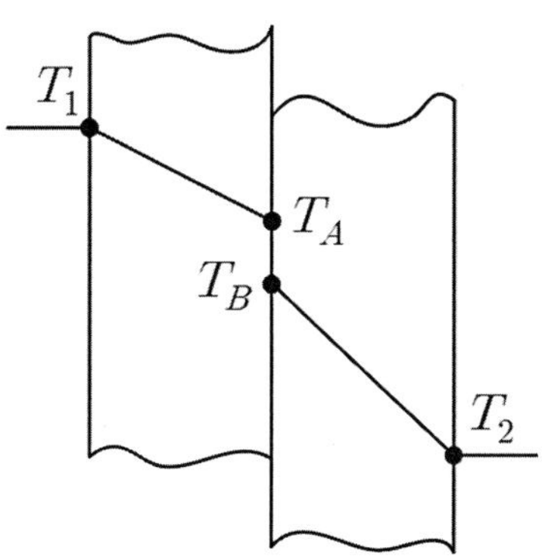

Fig. 4.7 Temperature drop at the interface between solids causing a contact resistance.

○ Heat balance

The concept of thermal resistance may work well and should be efficient in solving the complex network of thermal problems. But of course there are also many situations where the method cannot be applied directly. In this case, we must resort to the original governing equations or derive separate equations under relevant assumptions for simplification. In this case, we must consider heat balance, which means for steady state that

heat transfer in = heat transfer out (4.10)

Example

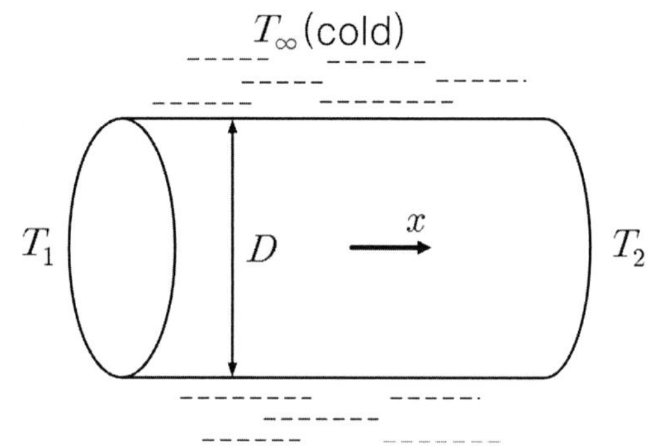

Fig. 4.8 1-D example of heat transfer between a cylindrical hot bar and the surrounding cold fluid.

Consider a bar submerged in a fluid maintained at $T = T_\infty$ (Fig. 4.8). Each end of the bar is maintained at a constant temperature. Consider a small cylindrical bar element as shown in Fig. 4.9.

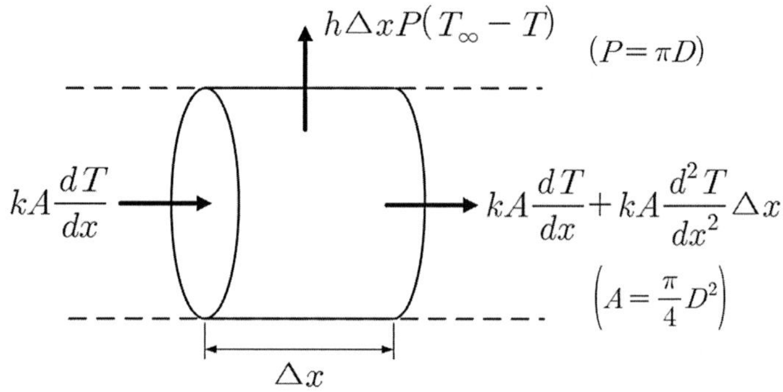

Fig. 4.9 Heat flux through the surfaces of an infinitesimal bar element.

The heat balance leads to

$$kA\frac{dT}{dx} = h\Delta x\pi D(T_\infty - T) + kA\frac{dT}{dx} + kA\frac{d^2T}{dx^2}\Delta x \tag{4.11}$$

It is assumed that there is no variation in the temperature along the radial direction in the bar, i.e., $T = T(x)$ only. Putting $T - T_\infty = \theta$ and $m^2 = 4h/kD$, we write

$$\theta'' - m^2\theta = 0$$

solution of which reads

$$\theta = c_1 e^{mx} + c_2 e^{-mx} \tag{4.12}$$

The boundary conditions at both ends of the bar can be used in fixing c_1 and c_2.

Homework

(1) Two solid slabs are in contact with each other (Fig. 4.10). The left-hand side is in contact with fluid at $T_{\infty 1}$ and the right-hand side with fluid at $T_{\infty 2}$. The thermal contact resistance between the two solids is R_{tc}. Derive the formula calculating $T_{s,1}$, T_A, T_B and $T_{s,2}$.

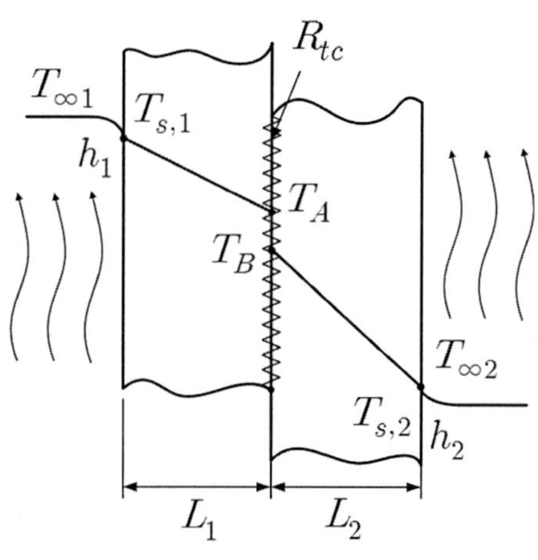

Fig. 4.10 One-dimensional model of heat transfer through two contacting slabs surrounded by fluids.

(2) A rod of square cross-section is attached to a hot wall maintained at $T = T_w$. The part of length L_1 near the wall is insulated (no heat transfer through the outer surface of the bar) and the other part of length L_2 is open to the air at T_∞ (Fig. 4.11). The rod end is also insulated. Derive the formula for the temperature distribution $T(x)$.

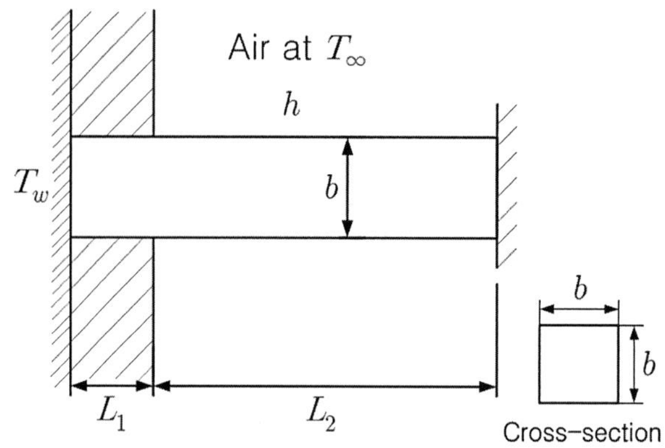

Fig. 4.11 A bar with a square cross-section attached to walls at both ends. The side surfaces are subject to two kinds of boundary conditions (refer to the text for details).

4.4 Two-Dimensional Steady-State Conduction

For a constant thermal conductivity, the conduction equation reads

$$\nabla^2 T = 0 \tag{4.13}$$

Appropriate boundary conditions, such as constant temperature, constant heat flux, constant convection coefficient with a surrounding fluid, etc, can be applied to solve the equation. Analytic or semi-analytic methods, such as separation of variables $T = T_1(x)T_2(y)$, can be used to solve the equation, but the method is limited to very simple geometries. In most practical cases, numerical methods, e.g., finite difference, finite volume, or finite element can be used.

4.5 Transient Conduction

4.5.1 Lumped capacitance method

On principle, we may use the full conduction equation

$$\frac{\partial T}{\partial t} = \alpha \nabla^2 T + \frac{\alpha}{k}\dot{q} \tag{4.14}$$

to solve a transient conduction problem. But in order to capture rough estimate for variables involved in the problem, we can also apply simplified concepts and related equation. In the lumped capacitance model, the temperature within a given object is assumed to be spatially uniform in distribution and only the temporal variation is pursued.

Suppose a solid is submerged in a cold fluid at T_∞ (Fig. 4.12). The heat transfer from the hot solid to the fluid is

$$q = hA_s(T - T_\infty)$$

where A_s is the outer surface area of the solid. Such heat loss results in temperature drop in the solid. The rate of heat loss in the solid is $\rho c_p V(dT/dt)$. Thus, we require (V is volume)

$$\rho c_p V\frac{dT}{dt} = -hA_s(T - T_\infty) \tag{4.15}$$

Applying the initial condition, $T = T_i$ at $t = 0$, gives

$$T = T_\infty + (T_i - T_\infty)\exp(-mt) \tag{4.16}$$

where $m = hA_s/(\rho c_p V)$.

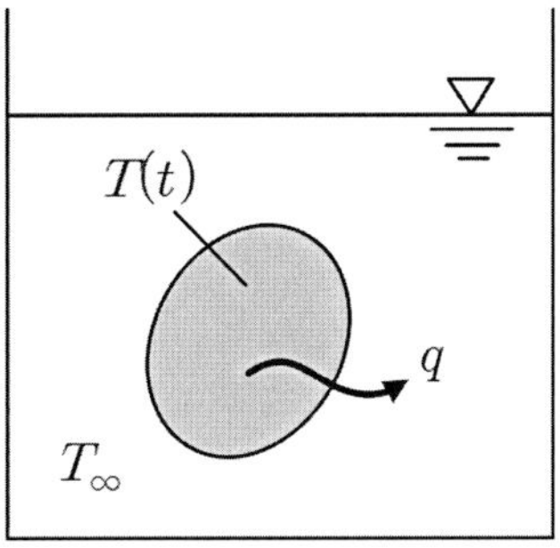

Fig. 4.12 Transient heat-transfer problem between a solid and the surrounding fluid.

4.5.2 One-dimensional transient problem

The governing equation is, for $T = T(t, x)$,

$$\frac{\partial T}{\partial t} = \alpha\frac{\partial^2 T}{\partial x^2} + \frac{\alpha}{k}\dot{q} \tag{4.17}$$

Let us assume that there is no heat source, $\dot{q} = 0$, and T can be expressed in a separation-of-variable form, $T = \Theta(t)X(x)$. Then

$$\Theta' X = \alpha\Theta X'' \tag{4.18}$$

Dividing both sides by $\alpha\Theta X$ gives

$$\frac{\Theta'}{\alpha\Theta} = \frac{X''}{X} = \text{constant} = -\lambda^2 < 0$$

We set the constant as negative in order to avoid exponential growth of Θ in time.

Then,

$$\Theta = \exp(-\alpha\lambda^2 t)$$
$$X = C_1 \sin(\lambda x) + C_2 \cos(\lambda x)$$

and

$$T = \exp(-\alpha\lambda^2 t)[C_1 \sin(\lambda x) + C_2 \cos(\lambda x)] \tag{4.19}$$

Details of the solution depend on the boundary conditions as well as the initial condition. If, coincidentally, the boundary condition is $T=0$ at $x=0$ and L, and the initial condition is $T=\sin(\pi x/L)$, then we derive $C_2=0$ and $C_1=1$, giving

$$T = \exp\left[-\alpha\left(\frac{\pi}{L}\right)^2 t\right]\sin\left(\frac{\pi}{L}x\right) \tag{4.20}$$

as the exact solution. But such conditions are very rare.

To tackle the practical problems, we must consider sum of all the modes of T with possible values of λ as follows

$$T = \sum_{n=1}^{\infty} \exp(-\alpha\lambda_n^2 t)[C_{1n}\sin(\lambda_n x) + C_{2n}\cos(\lambda_n x)] \tag{4.21}$$

Example

As a concrete example, consider a solid slab with width L maintained initially at $T=0$. The left-hand side end of the slab is then attached to a hot heat source maintained at $T=T_1$ while the other end at $T=0$. Since initially $T=0$, we cannot use the original series form for the solution T. Instead, we must consider perturbation from the steady-state solution $T_s = T_1(1-x/L)$. Thus, we write

$$T = T_1\left(1-\frac{x}{L}\right) + \sum_{n=1}^{\infty}\exp(-\alpha\lambda_n^2 t)[C_{1n}\sin(\lambda_n x)]$$

We set $C_{2n}=0$ because the boundary condition is now $T=0$ at $x=1$ and $x=L$. We also require $\lambda_n = n\pi/L$ to satisfy the boundary condition. We determine C_{1n} from the initial condition;

$$0 = T_1\left(1-\frac{x}{L}\right) + \sum_{n=1}^{\infty} C_{1n}\sin\left(\frac{n\pi}{L}x\right)$$

which corresponds to the Fourier-expansion problem. C_{1n} is given from

$$C_{1n} = -\frac{T_1}{L}\int_0^{2L}\left(1-\frac{x}{L}\right)\sin\left(\frac{n\pi}{L}x\right)dx$$

Figure 4.13 shows qualitative feature of the temperature variation in time in the slab.

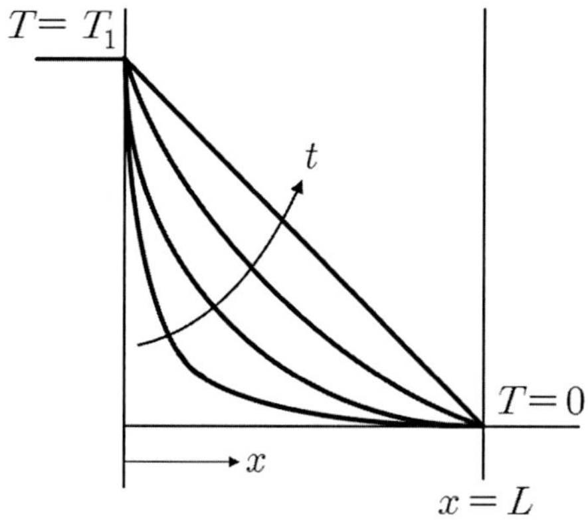

Fig. 4.13 Time history of temperature variation in a slab with both ends being maintained at constant temperatures.

Homework

Obtain the solution $T(t,x)$ for the case of insulation ($\partial T/\partial x = 0$) at $x = L$ instead of $T = 0$ there.

4.6 Convective Heat Transfer

4.6.1 Convective heat transfer coefficient

The ultimate goal of this section is to derive the formula for the convective heat transfer coefficient h. In terms of h, we can write the heat transfer rate as

$$q = hA(T_s - T_\infty) \tag{4.22}$$

But this applies only to the case in which h is constant on A. When h varies on the surface, we must use

$$q = (T_s - T_\infty)\int h\,dA \tag{4.23}$$

under further restriction of constant $(T_s - T_\infty)$. Thus, we can define the averaged coefficient $\bar{h}$;

$$\bar{h} = \frac{1}{A}\int h\,dA \tag{4.24}$$

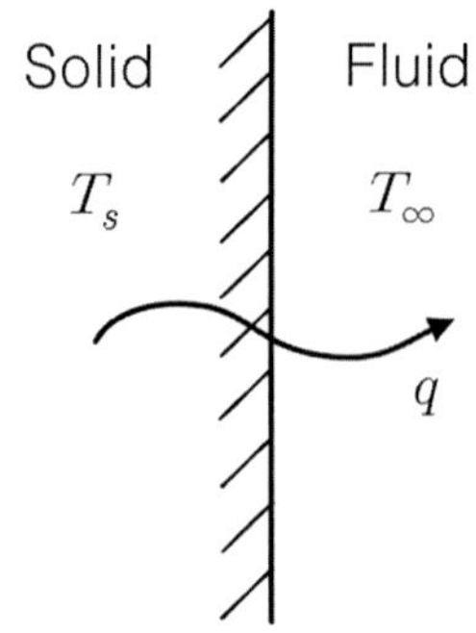

Fig. 4.14 Heat flux q across the interface between a solid and fluid in motion.

In fact, the heat transfer between a solid and fluid in motion (Fig. 4.14) is determined by the conduction equation (k_f: the thermal conductivity of fluid)

$$q = k_f A \left[\frac{\partial T}{\partial n}\right]_{n=0}$$

where n is the coordinate normal to the surface. As an example, consider a parabolic profile for T within a channel (Fig. 4.15),

$$T = \frac{4}{H^2}(T_m - T_w)\,y(H-y) + T_w$$

Then

$$\left[\frac{\partial T}{\partial n}\right]_w = \left[\frac{\partial T}{\partial y}\right]_w = \frac{4}{H}(T_m - T_w)$$

So, we get

$$h = 4\frac{k_f}{H}$$

from $q \equiv hA(T_m - T_w)$.

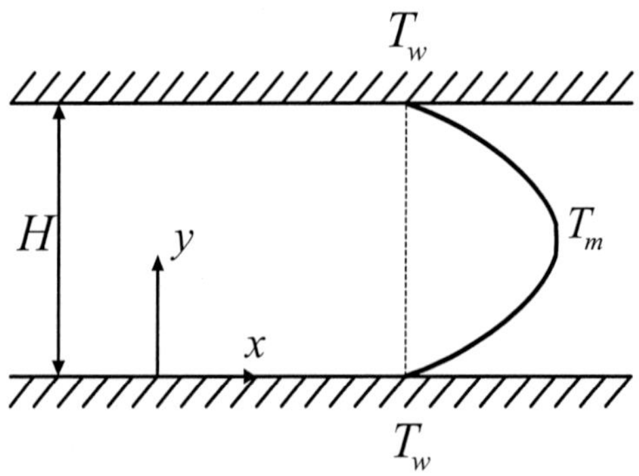

Fig. 4.15 Parabolic profile of fluid temperature within a channel.

Consider a temperature distribution, which is sharp within the thin layer near the wall (Fig. 4.16),

$$T = T_m - (T_m - T_w)\exp\left(-\frac{y}{\delta}\right)$$

Then,

$$\left[\frac{\partial T}{\partial y}\right]_w = \frac{1}{\delta}(T_m - T_w)$$

From the relation

$$q = hA(T_m - T_w) = k_f A\left[\frac{\partial T}{\partial y}\right]_w$$

we get

$$h = \frac{k_f}{\delta}$$

This implies that the heat transfer coefficient becomes larger when the layer is thinner.

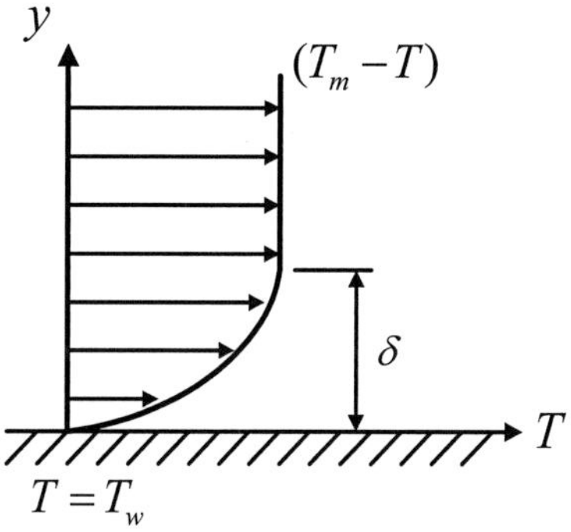

Fig. 4.16 Profile of temperature in the fluid in contact with a solid at $y=0$, showing a sharp variation with the distance y.

The above example shows that the heat transfer coefficient depends on the specific profile of temperature as a function of the coordinate normal to the wall. As will be shown below, the functional form of the temperature is dependent on the convective effect of the fluid, and a strong fluid motion in general leads to a steep profile in the distribution of T, which in turn implies an enhanced heat transfer through increased h value.

4.6.2 Derivation of energy equation

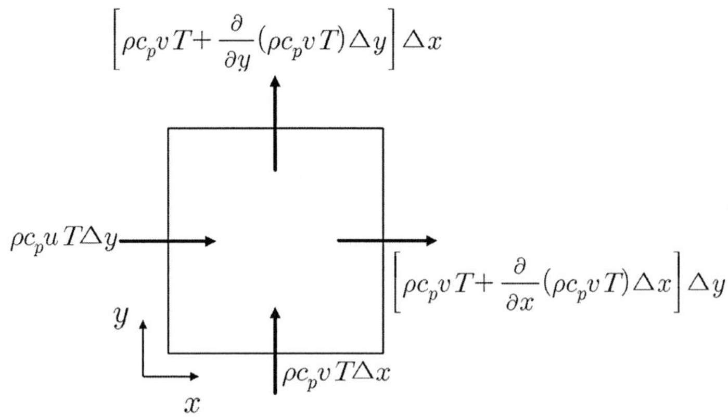

Fig. 4.17 Convective heat transfer through surfaces of an infinitesimal fluid element.

Consider a small fluid element with side lengths, Δx and Δy. The conductive heat transfer rate was derived preciously;

$$q^k_{net} = q^k_x + q^k_y - q^k_{x+\Delta x} - q^k_{y+\Delta y} = \left[\frac{\partial}{\partial x}\left(k\frac{\partial T}{\partial x}\right) + \frac{\partial}{\partial y}\left(k\frac{\partial T}{\partial y}\right)\right]\Delta x \Delta y \qquad (4.25)$$

Now, we derive the net heat transfer rate due to the fluid motion (Fig. 4.17). The heat transfer rate through the left-hand side boundary is

$$q^c_x = \rho c_p u T \Delta y$$

and that through the right-hand side boundary is

$$q^c_{x+\Delta x} = \left[\rho c_p u T + \frac{\partial}{\partial x}(\rho c_p u T)\Delta x\right]\Delta y$$

Similarly,

$$q^c_y = \rho c_p v T \Delta x$$

$$q^c_{y+\Delta y} = \left[\rho c_p v T + \frac{\partial}{\partial y}(\rho c_p v T)\Delta y\right]\Delta x$$

The net heat transfer rate becomes

$$q_{net} = q^k_{net} + q^c_{net} = q^k_x + q^k_y - q^k_{x+\Delta x} - q^k_{y+\Delta y} + q^c_x + q^c_y - q^c_{x+\Delta x} - q^c_{y+\Delta y}$$

Equating this with the transient term $\rho c_p(\partial T/\partial t)\Delta x \Delta y$ results in

$$\frac{\partial T}{\partial t} + \frac{\partial (uT)}{\partial x} + \frac{\partial (vT)}{\partial y} = \alpha \nabla^2 T + \frac{\alpha}{k}\dot{q}$$

Applying the continuity equation

$$\frac{\partial u}{\partial x} + \frac{\partial v}{\partial y} = 0$$

we have

$$\frac{\partial T}{\partial t}+u\frac{\partial T}{\partial x}+v\frac{\partial T}{\partial y}=\alpha\nabla^2 T+\frac{\alpha}{k}\dot{q} \tag{4.26}$$

The vector form;

$$\frac{\partial T}{\partial t}+(\mathbf{u}\cdot\nabla)T=\alpha\nabla^2 T+\frac{\alpha}{k}\dot{q} \tag{4.27}$$

For a very viscous fluid, heat is generated due to dissipation;

$$\frac{\partial T}{\partial t}+(\mathbf{u}\cdot\nabla)T=\alpha\nabla^2 T+\frac{\mu}{\rho c_p}\left\{\left(\frac{\partial u}{\partial y}+\frac{\partial v}{\partial x}\right)^2+2\left[\left(\frac{\partial u}{\partial x}\right)^2+\left(\frac{\partial v}{\partial y}\right)^2\right]\right\}+\frac{\alpha}{k}\dot{q} \tag{4.28}$$

4.6.3 Boundary-layer equation

For the fluid with low viscosity or high velocity or with large scale of length (in fact with large Reynolds number), we expect thin layers, called boundary layers near walls (Fig. 4.18). In this case, we can neglect the diffusion term $\partial^2 u/\partial x^2$ compared with $\partial^2 u/\partial y^2$ in the Navier-Stokes equations; for steady state,

$$u\frac{\partial u}{\partial x}+v\frac{\partial u}{\partial y}=-\frac{1}{\rho}\frac{\partial p}{\partial x}+\nu\frac{\partial^2 u}{\partial y^2} \tag{4.29}$$

$$0=-\frac{1}{\rho}\frac{\partial p}{\partial y} \tag{4.30}$$

where x is coordinate along the wall and y is normal to x.

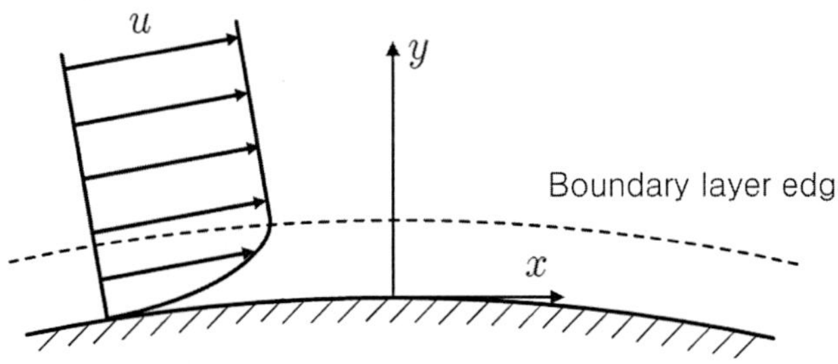

Fig. 4.18 Velocity profile within the boundary layer showing a sharp variation in the fluid velocity.

Similarly, we expect a thin thermal boundary layer when α is very small; for steady state,

$$u\frac{\partial T}{\partial x}+v\frac{\partial T}{\partial y}=\alpha\frac{\partial^2 T}{\partial y^2} \tag{4.31}$$

where $\dot{q}$ and dissipation are not considered.

4.7 External Flow

○ Flat plate in a parallel flow

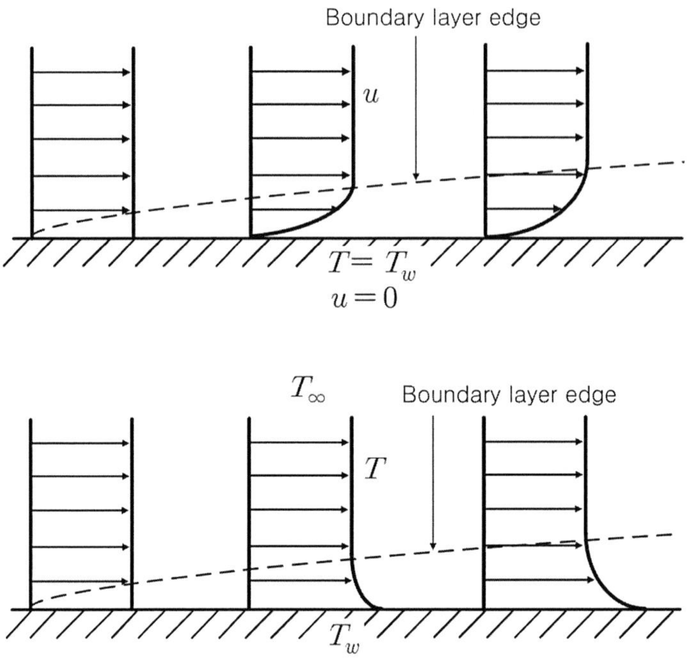

Fig. 4.19 Velocity (top) and temperature (bottom) profiles within the boundary layer over a flat plate subjected to a uniform parallel flow.

The boundary layer equations are:

$$u\frac{\partial u}{\partial x}+v\frac{\partial u}{\partial y}=\nu\frac{\partial^2 u}{\partial y^2} \tag{4.32}$$

$$u\frac{\partial T}{\partial x}+v\frac{\partial T}{\partial y}=\alpha\frac{\partial^2 T}{\partial y^2} \tag{4.33}$$

$$\frac{\partial u}{\partial x}+\frac{\partial v}{\partial y}=0 \tag{4.34}$$

The boundary conditions are:

$u=v=0$ at $y=0$, $u\rightarrow u_\infty$ as $y\rightarrow\infty$

$T=T_w$ at $y=0$, $T\rightarrow T_\infty$ as $y\rightarrow\infty$

Introduce similarity variable $\eta=y\sqrt{u_\infty/\nu x}$ and let $u=u_\infty f'(\eta)$. Then, we arrive at

$$f'''+\frac{1}{2}ff''=0 \tag{4.35}$$

with the boundary conditions,

$$f = f' = 0 \quad \text{at} \quad \eta = 0$$

$$f' \to 1 \quad \text{as} \quad \eta \to \infty$$

Numerical methods must be used to solve above equations.

The boundary-layer thickness δ is defined as

$$\frac{u}{u_\infty} = f'\left(\delta\sqrt{u_\infty/\nu x}\right) = 0.99$$

Numerical data give

$$\delta\sqrt{\frac{u_\infty}{\nu x}} = 5 \quad \text{or} \quad \delta = \frac{5}{\sqrt{u_\infty/\nu x}} \tag{4.36}$$

Figure 4.20 shows qualitative description of the displacement thickness δ as a function of x.

Fig. 4.20 Schematic showing the variation of the displacement thickness.

The shear stress τ_w is

$$\tau_w = \mu\frac{\partial u}{\partial y}\bigg|_{y=0} = \mu u_\infty\sqrt{\frac{u_\infty}{\nu x}}\,f''(0) = 0.332\,u_\infty\sqrt{\frac{\rho\mu u_\infty}{x}} \tag{4.37}$$

For the energy equation, we introduce $\theta = (T - T_w)/(T_\infty - T_w)$ so that $\theta = 0$ at $y = 0$ and $\theta \to 1$ as $y \to \infty$. Next, we assume $\theta = \theta(\eta)$. For the Prandtl number $\mathrm{Pr} = \nu/\alpha$, we have

$$\theta'' + \frac{1}{2}\mathrm{Pr}\, f\,\theta' = 0 \tag{4.38}$$

Numerical data show that

$$\theta'_0 = 0.332\,\mathrm{Pr}^{1/3}$$

for $\mathrm{Pr} \gtrsim 0.6$. From the definition of heat flux

$$q'' = k_f\left(\frac{\partial T}{\partial y}\right)_0 = h(T_\infty - T_w)$$

we get

$$h = \frac{k_f}{T_\infty - T_w}\left(\frac{\partial T}{\partial y}\right)_0 = \frac{k_f}{T_\infty - T_w}(T_\infty - T_w)\theta'_0\sqrt{\frac{u_\infty}{\nu x}} = 0.332\mathrm{Pr}^{1/3}\sqrt{\frac{u_\infty}{\nu x}}\,k_f$$

The Nusselt number (local) defined as $\mathrm{Nu}_x = hx/k_f$ is

$$\mathrm{Nu}_x = 0.332\,\mathrm{Pr}^{1/3}\,\mathrm{Re}_x^{1/2} \tag{4.39}$$

where $\mathrm{Re}_x = u_\infty x/\nu$ is the local Reynolds number.

○ Thickness of thermal boundary layer

It turns out that

$$\delta_t = \delta\,\mathrm{Pr}^{-1/3} \tag{4.40}$$

Thus, high thermal diffusivity α means low Pr and large thermal boundary layer, and vice versa for low α. Figure 4.21 reveals dramatic difference in the thermal and fluid boundary-layer thickness between the case with very small Prandtl number and that with very large Prandtl number.

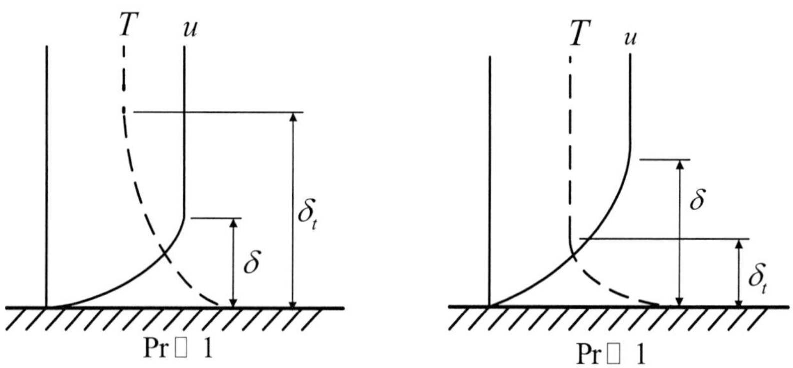

Fig. 4.21 Difference in the velocity (solid lines) and temperature (dashed lines) profiles between the case with very small (left) and very large (right) Prandtl numbers.

○ Dimensionless heat transfer coefficient

The Nusselt number $\mathrm{Nu} = hL/k$ can be written as a function of Reynolds number Re and Prandtl number Pr;

$$\mathrm{Nu} = C\,\mathrm{Re}^m\,\mathrm{Pr}^n \tag{4.41}$$

For the flat-plate case, $m = 1/2$ and $n = 1/3$. In practical thermal problems, C, m and n are determined from CFD or experiment.

4.8 Internal Flow

○ Developing flow and heat transfer

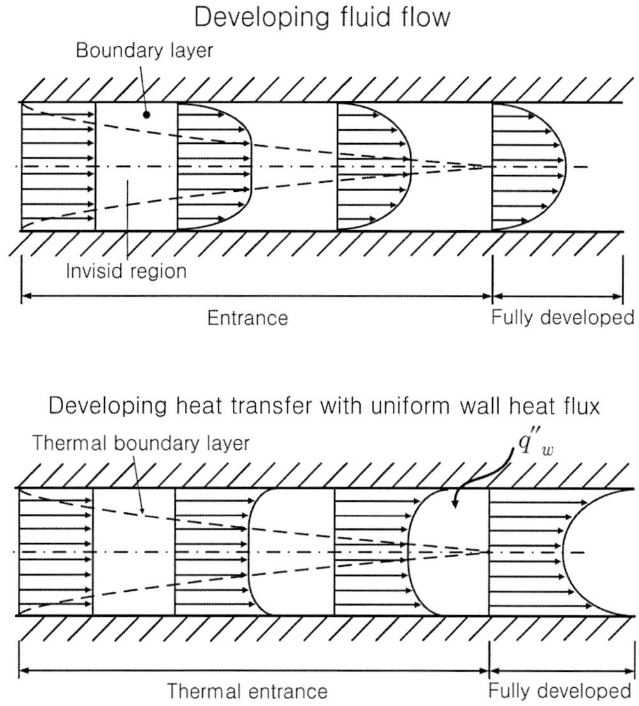

Fig. 4.22 Developing fluid flow (top) and developing heat transfer (bottom).

○ Heat balance

Consider one-dimensional heat transfer through a duct with an arbitrary cross-section as shown in Fig. 4.23.

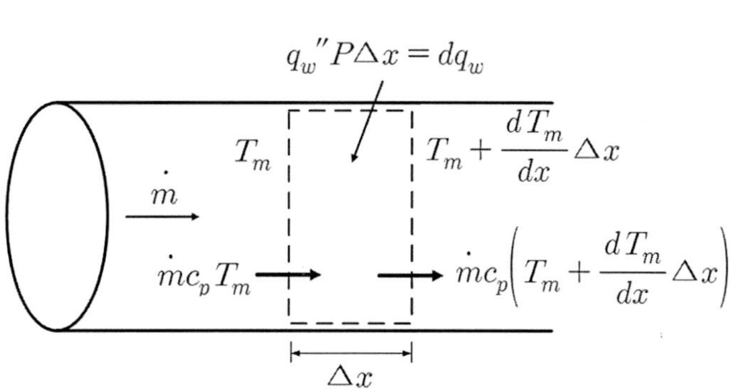

Fig. 4.23 A small element inside a duct showing heat transfer through the surrounding surfaces.

Heat flux through the input and output surfaces of the small fluid element of length Δx (Fig. 4.23) can be obtained as follows.

Input : $dq_w + \dot{m} c_p T_m$

Output : $\dot{m} c_p \left(T_m + \frac{dT_m}{dx} \Delta x \right)$

They must show balance with the heat flux through the side surface.

$$\therefore \quad \frac{dT_m}{dx} = \frac{1}{\dot{m} c_p} \frac{dq_w}{\Delta x} = \frac{q''_w P}{\dot{m} c_p} \tag{4.42}$$

- Constant surface heat flux: $q''_w = \text{const}$

$$T_m = \frac{q''_w P}{\dot{m} c_p} x + T_{m,i} \tag{4.43}$$

where $T_{m,i}$ is the inlet mean temperature at $x = 0$. So, T_m is a linear function of x as shown in Fig. 4.24.

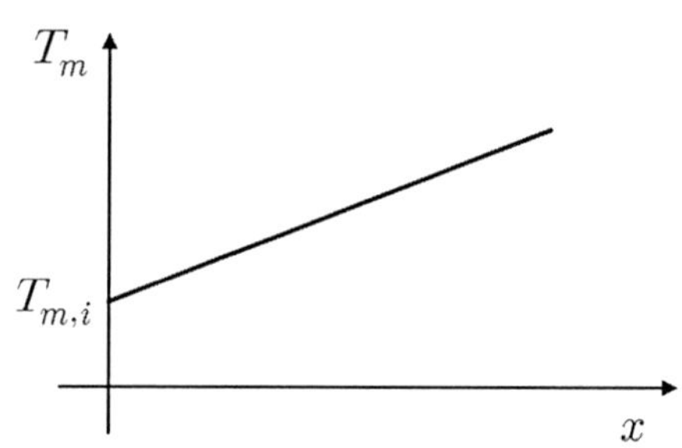

Fig. 4.24 Linear variation of T_m along the duct for a constant surface-heat-flux.

- Constant surface temperature: $T_w = \text{const}$

$$q'' = h(T_w - T_m)$$

$$\frac{dT_m}{dx} = \frac{hP}{\dot{m} c_p}(T_w - T_m)$$

Put $\theta = (T_w - T_m)/(T_w - T_{m,i})$, then

$$\frac{d\theta}{dx} + \frac{hP}{\dot{m} c_p}\theta = 0$$

$$\theta = \exp\left(-\frac{hP}{\dot{m} c_p} x \right)$$

or

$$T_m = T_w - (T_w - T_{m,i}) \exp\left(-\frac{hP}{\dot{m} c_p} x \right) \tag{4.44}$$

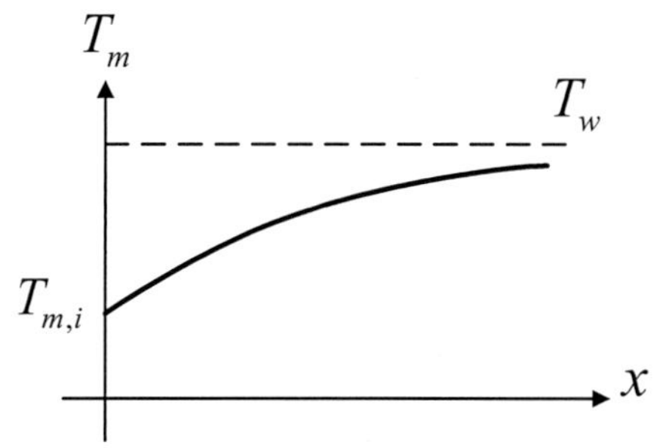

Fig. 4.25 Non-linear variation of T_m along the duct for a constant surface temperature, T_w.

Homework

(1) Consider a fully developed flow with $u = \text{const}$ through a channel between two parallel plates separated by $2H$ (Fig. 4.26). Heat flux through each of the walls is set constant, i.e.,

$$q''_w = k\left(\frac{\partial T}{\partial y}\right)_w = \text{const}$$

Starting with the energy equation

$$u\frac{\partial T}{\partial x} + v\frac{\partial T}{\partial y} = \alpha\left(\frac{\partial^2 T}{\partial x^2} + \frac{\partial^2 T}{\partial y^2}\right)$$

and with the assumption, $T = T_b(x) + \theta(y)$, solve the problem to get the solution $T(x,y)$.

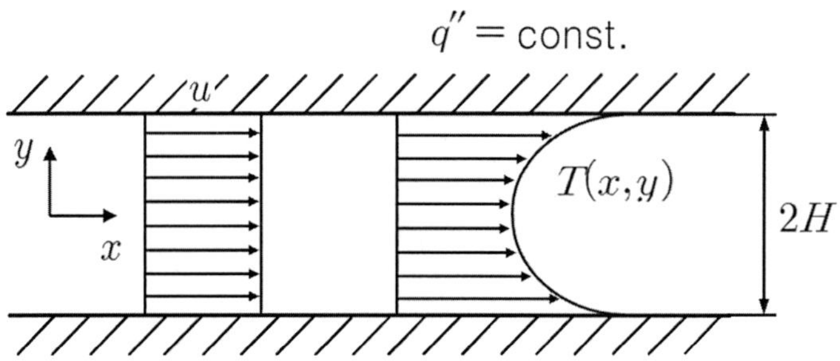

Fig. 4.26 Temperature profile in a duct for a uniform velocity profile with a constant heat flux through the walls.

(2) The heat transfer coefficient for a flat plate was derived as

$$h = 0.332\,\text{Pr}^{1/3} k_f \sqrt{\frac{u_\infty}{\nu x}}$$

Obtain the averaged coefficient $\bar{h}$ over the interval $0 \le x \le L$.

(3) The mean temperature $T_m(x)$ within the channel was found to be

$$T_m(x) = T_w - (T_w - T_{m,i})\exp\left(-\frac{hP}{\dot{m}c_p}x\right)$$

Consider various cross-sections (i.e., circle, square, rectangle, triangle) with the same cross-section area A, and discuss which type would give the fastest adjustment of T_m to T_w.

4.9 Free Convection

The 2-dimensional incompressible, steady-state equations for fluid flow are:

$$\rho(\mathrm{u}\cdot\nabla)\mathrm{u} = -\nabla p + \mu\nabla^2\mathrm{u} + \mathrm{f} \tag{4.45}$$

where f represents a body force. Under the effect of gravitation, f reads $\mathrm{f} = \rho\mathrm{g} = -\rho g\mathrm{i}$. If $\rho = \mathrm{const}$ everywhere in the domain, it contributes to the pressure, and we can replace p by a new pressure $\tilde{p} = p + \rho g x$ so that the equations convert to

$$\rho(\mathrm{u}\cdot\nabla)\mathrm{u} = -\nabla\tilde{p} + \mu\nabla^2\mathrm{u} \tag{4.46}$$

implying that no body force is existent. But, when there occurs a small variation of density, due to the temperature difference, in the space, the problem is not so simple.

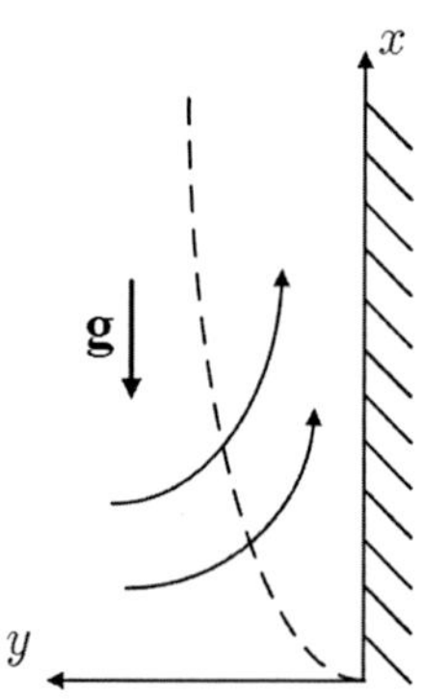

Fig. 4.27 Natural (free) convection near a vertical wall.

Let $\rho = \rho_\infty - \Delta\rho$ where ρ_∞ denotes the fluid density far away from the domain of interest (such as boundary layers) where temperature is maintained at $T_\infty = \mathrm{const}$ over a wide region. We further assume that $\Delta\rho$ is linearly proportional to the temperature difference, i.e., $\Delta\rho = \rho\beta(T - T_\infty)$ where β is the thermal expansion coefficient. Referred to as Boussinesq approximation, the equations read

$$\begin{aligned}\rho(\mathrm{u}\cdot\nabla)\mathrm{u} &= -\nabla p + \mu\nabla^2\mathrm{u} - \rho_\infty g\mathrm{i} + \rho g\beta(T - T_\infty)\mathrm{i} \\ &= -\nabla\tilde{p} + \mu\nabla^2\mathrm{u} + \rho g\beta(T - T_\infty)\mathrm{i}\end{aligned} \tag{4.47}$$

The last term on the right-hand side cannot be written as a gradient of a scalar function, implying that this equation must be solved in coupling with the energy equation

$$(\mathrm{u}\cdot\nabla)T=\alpha\nabla^2 T \tag{4.48}$$

○ Laminar free convection on a vertical plate

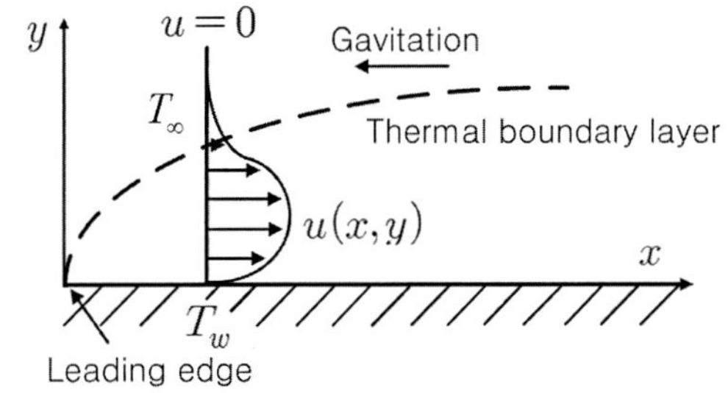

Fig. 4.28 Velocity profile given for a free convection problem near the wall $y=0$ (practically 'vertical' wall).

Boundary-layer approximation gives

$$\frac{\partial u}{\partial x}+\frac{\partial v}{\partial y}=0 \tag{4.49}$$

$$u\frac{\partial u}{\partial x}+v\frac{\partial u}{\partial y}=g\beta(T-T_\infty)+\nu\frac{\partial^2 u}{\partial y^2} \tag{4.50}$$

$$u\frac{\partial T}{\partial x}+v\frac{\partial T}{\partial y}=\alpha\frac{\partial^2 T}{\partial y^2} \tag{4.51}$$

Introduce similarity variables,

$$\eta=\frac{y}{x}\left(\frac{Gr_x}{4}\right)^{1/4} \tag{4.52}$$

where

$$Gr_x=\frac{g\beta(T_w-T_\infty)x^3}{\nu^2} \quad \text{(local Grashof number)} \tag{4.53}$$

and let

$$u=\frac{2\nu}{x}Gr_x^{1/2}f'(\eta),\ \ \theta=\frac{T-T_\infty}{T_w-T_\infty} \tag{4.54}$$

to get

$$f'''+3ff''-2(f')^2+\theta=0 \tag{4.55}$$

$$\theta''+3\,\mathrm{Pr}\,f\theta'=0 \tag{4.56}$$

Boundary conditions; $f=f'=0$, $\theta=1$ at $\eta=0$

$f'=0$, $\theta\rightarrow 0$ as $\eta\rightarrow\infty$

Homework

(1) Derive the equations for $f(\eta)$ and $\theta(\eta)$ and the boundary conditions.

(2) Derive the equation for Nu_x and Nu_L (below).

From the numerical solutions of those equations, we can derive the dimensionless heat transfer coefficient, Nusselt number. The heat flux is

$$g''_w = -k\left(\frac{\partial T}{\partial y}\right)_0 = -\frac{k}{x}(T_w - T_\infty)\left(\frac{Gr_x}{4}\right)^{1/4}\left(\frac{d\theta}{d\eta}\right)_0 \equiv h(T_w - T_\infty) \qquad (4.57)$$

Let $(d\theta/d\eta)_0 = -g(\mathrm{Pr})$ as a function of Pr. Then,

$$\mathrm{Nu}_x \equiv \frac{hx}{k} = \left(\frac{Gr_x}{4}\right)^{1/4} g(\mathrm{Pr}) \qquad (4.58)$$

For a wide range of Pr, the correlation

$$g(\mathrm{Pr}) = \frac{0.75\,\mathrm{Pr}^{1/2}}{(0.609 + 1.221\,\mathrm{Pr}^{1/2} + 1.238\,\mathrm{Pr})^{1/4}} \qquad (4.59)$$

gives very accurate results (within 0.5%).

○ Various geometries under free convection

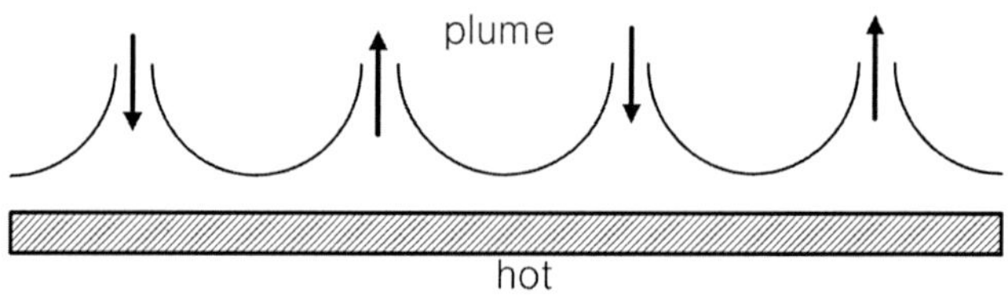

(a)

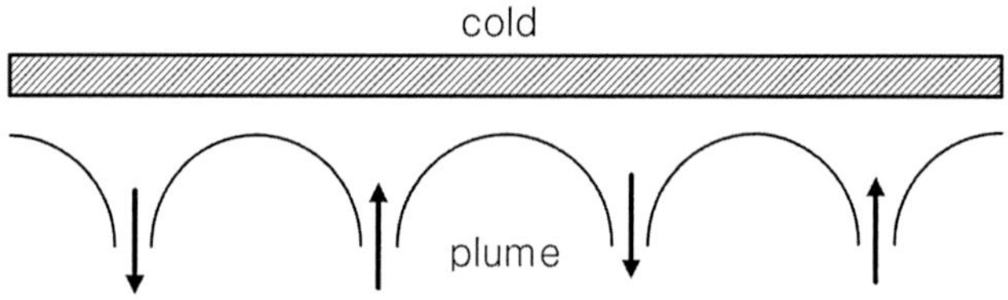

(b)

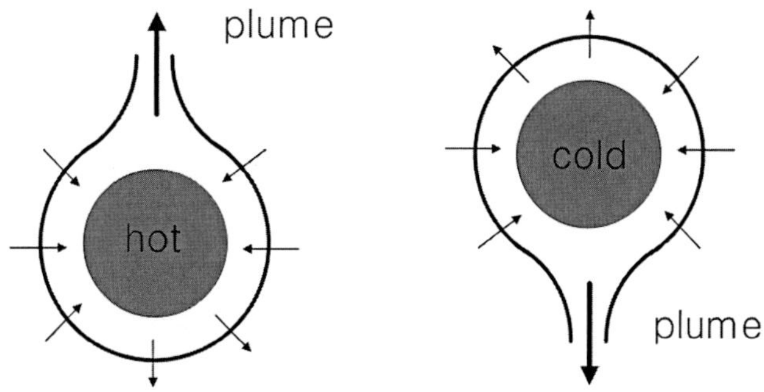

(c)

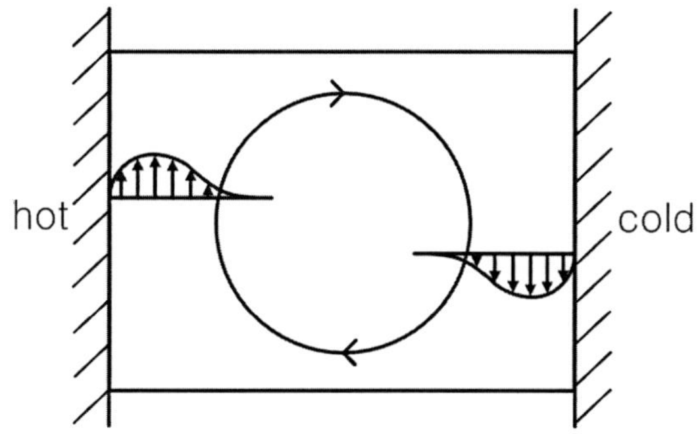

(d)

Fig. 4.29 Various configurations of free convection problems of interest. Upward- and downward-flow near the (a) hot and (b) cold horizontal plates, (c) plumes around a solitary obstacle, and (d) circulatory flow inside a cavity.

Chapter 5. General Chemistry

Atomic elements

Al	aluminium	Li	lithium
Sb	antimony	Mg	magnesium
Ar	argon	Mn	manganese
Ba	barium	Hg	mercury
Be	beryllium	Mo	molybdenum
Bi	bismuth	Ne	neon
B	boron	Ni	nickel
Br	bromine	N	nitrogen
Cd	cadmium	O	oxygen
Ca	calcium	P	phosphorus
C	carbon	Pt	platinum
Ce	cerium	K	potassium
Cs	cesium	Ra	radium
Cl	chlorine	Rn	radon
Cr	chromium	Rh	rhodium
Co	cobalt	Rb	rubidium
Cu	copper	Sc	scandium
F	fluorine	Se	selenium
Ga	gallium	Si	silicon
Ge	germanium	Ag	silver
Au	gold	Na	sodium
He	helium	S	sulfur
H	hydrogen	Sn	tin
I	iodine	Ti	titanium
Ir	iridium	W	tungsten
Fe	iron	U	uranium
Kr	krypton	V	vanadium
Pb	lead	Xe	xenon
		Zn	zinc

5.1 Some Properties of the Elements

5.1.1 Elements, compounds and mixtures

○ elements; substances that cannot be decomposed into simpler materials by chemical reactions.

element	Sodium	Potassium	Iron	Copper	Silver	Gold
symbol	Na	K	Fe	Cu	Ag	Au
Latin name	Natrium	Kalium	Ferrum	Cuprum	Argentum	Aurum

○ compound(화합물); substance formed from two or more different elements in which the elements are always combined in the same fixed (i.e. constant) proportions by mass. Ex.; H_2O, NaCl

○ mixture(혼합물); Ex.; CO_2 mixed in H_2O
solution; homogeneous mixture. Ex.; thoroughly stirred mixture of sugar in water.
Ex.; brass. Ex.; air

5.1.2 Symbols, formulas and equations

○ diatomic molecules; Ex.; N_2, O_2, H_2

○ hydrates; compounds whose crystals contain water molecules in fixed ratios. Ex.; $CaSO_4$ - $2H_2O$.

○ chemical equations; Ex.; Zn + S ----> ZnS
(zinc) (sulfur) (zinc sulfate)
↳ reactants ↳ product

5.1.3 The structure of matter: atoms and subatomic

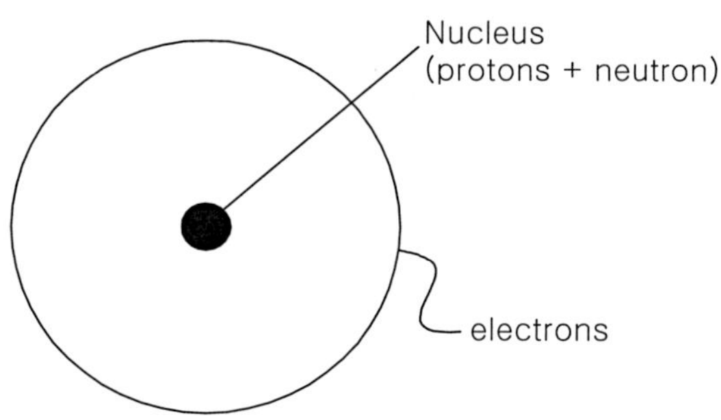

Fig. 5.1 Structure of atom.

○ atomic number (Z) = number of protons
mass number (A) = number of protons + number of neutrons

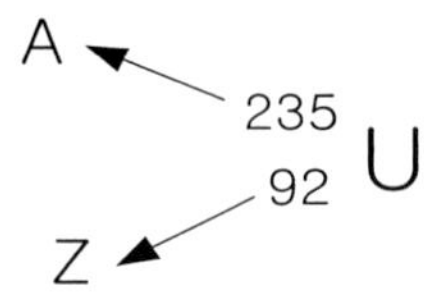

○ atomic mass = (average) mass of one atom; the relative magnitude is "u" [atomic mass unit, sometimes denoted by Da (Dalton)] corresponding to exactly 1/12 of the mass of the carbon atom C-12, which has the molar mass 1.66054×10^{-24} g; e.g., atomic mass of H is 1.0078 u in terms of the atomic mass unit or 1.3946×10^{-25} g in terms of gram.

5.1.4 Reactions of the elements: Formation of molecular and ionic compounds

○ molecular compound (분자화합물): compound composed of molecules. Ex.; H_2O
○ ionic compound (이온 화합물): compound composed of ions.

Ex.; <u>Na</u> + <u>Cl</u> ---> Na^+ + Cl^-
↳e- ↗

5.1.5 Ionic compounds and their properties

○ cation; positively charged ion. Ex.; Na^+
anion; negatively charged ion. Ex.; Cl^-

○ transition material; it is able to form more than one positive ion.

chromium	Cr^{2+}, Cr^{3+}
manganese	Mn^{2+}, Mn^{3+}
iron	Fe^{2+}, Fe^{3+}
cobalt	Co^{2+}, Co^{3+}
copper	Cu^{+}, Cu^{2+}
gold	Au^{+}, Au^{3+}

○ Formulas and names of some polyatomic ions

NH_4^+	ammonium ion	CO_3^{2-}	carbonate ion
H_3O^+	hydromium	HCO_3^-	hydrocarbonate
OH^-	hydroxide	SO_3^-	sulfite
CN^-	cynide	HSO_3^-	hydrogen sulfite
NO_2^-	nitride	SO_4^{2-}	sulfate

NO_3^- nitrate
ClO^- hypochlorite
ClO_2^- chlorite
ClO_3^- chlorate
MnO_4^- permanganate
HSO_4^- hydrogen sulfate
CrO_4^{2-} chromate
$Cr_2O_7^{2-}$ dichromate
PO_4^{3-} phosphate
$C_2H_3O_2^-$ acetate

○ Electrical properties.

pure water, solid NaCl ---> not conductors
water + salt, molten NaCl ---> conductors

5.1.6 Molecular compounds and their properties

○ compounds of carbon: the number and complexity of carbon compounds is enormous, and their study constitutes the major specialty called "organic chemistry".

some hydrocarbons	alcohols
CH_4 methane	CH_3OH methyl alcohol (methanol)
C_2H_6 ethane	C_2H_5OH ethyl alcohol (ethanol)
C_3H_8 propane	
C_4H_{10} butane	

5.1.7 Inorganic chemical nomenclature

○ binary compounds containing a metal and a nonmetal name; metal + ~ide

Ex.; CaO calcium oxide
ZnS zinc sulfide
Mg_3N_2 magnesium nitride

H^- hydride	C^{4-} carbide	Si^{4-} silicide
N^{3-} nitride	P^{3-} phosphide	Br^- bromide
O^{2-} oxide	S^{2-} sulfide	I^- iodide
F^- fluoride	Cl^- chloride	

○ binary compounds between two nonmetals

mono-- = 1 (often omitted)	hexa-- = 6
di-- = 2	hepta-- = 7
tri-- = 3	octa-- = 8
tetra-- = 4	nona-- = 9
penta-- = 5	deca-- = 10

Ex.; NO_2; nitrogen dioxide
N_2O_4; dinitrogen tetraoxide

HCl; hydrogen chloride
CO; carbon monoxide

○ binary acids and their salts

aqueous solutions	gas state
HCl; hydrochloric acid	hydrogen chloride
H_2S; hydrosulfuric acid	hydrogen sulfide

○ oxoacids and their salts

H_2SO_4; sulfuric acid	
H_2SO_3; sulfurous acid	SO_3^-; sulfite ion
HNO_3; nitric acid	NO_3^-; nitrate ion
HNO_2; nitrous acid	
$HClO$; hypochlorous acid	ClO^-; hypochlorite ion
$HClO_2$; chlorous acid	
$HClO_3$; chloric acid	
$HClO_4$; perchloric acid	ClO_4^-; perchlorate ion

5.2 Stoichiometry: Quantitative Chemical Relationships

5.2.1 Measuring moles of elements and compounds

○ 1 mol of carbon-12 = 12 g
↳ element

○ 1 mol of $Ca_3(PO_4)_2$ = 3×40.08 + 2×30.97 + 8×16 = 310.18 g (molar mass)
↳ atomic mass of Ca

5.2.2 Percentage composition

○ percentage by mass of element $= \dfrac{\text{mass of elements}}{\text{mass of whole sample}} \times 100$

5.2.3 Reaction in solution

○ solution(용액); a homogeneous mixture in which the molecules or ions of the components are fully intermingled.
○ solvent(용매); medium into which the solutes are mixed or dissolved (water).
○ solute(용질); any substance dissolved in the solvent.
○ dilute solution; the ratio of solute to solvent is small, sometimes very small.

○ concentrated solution; the ratio of solute to solvent is large. Ex.; syrup
○ saturated solution(포화용액); the situation of a solution in which addition of solute does not give additional dissolution.
○ solubility(용해도); usually the number of grams of solute that dissolve in 100g of solvent at a given temperature.
○ supersaturated solution(과포화 용액); a solution that actually contains more solute than required for saturation at a given temperature.
○ percipitate(침전); when a reaction is carried out in a solution, one of the products that forms has a low solubility in the solvent. As this substance forms, it separates from the solution as a solid, which we call a precipitate.

5.2.4 Molar concentration

○ molar concentration (molarity; ***M***, 몰농도) $= \dfrac{\text{mole of mass}}{\text{liters of solution}}$

Ex,; 0.1 mol of NaCl in 1.00L has a molarity of 0.100 ***M*** NaCl

5.3 Reactions between Ions in Aqueous Solutions

5.3.1 Electrolytes and non-electrolytes

○ electrolytes(전해질); solutes such as $CuSO_4$ or NaCl, which yield electrically conducting aqueous solutions, are called electrolytes.
○ dissociation(해리); separation of ions from each other existing as independent particles that are surrounded by molecules of the solvent.
○ strong electrolyte; electrically conducting solutions in which no undissociated solutes exist.
○ equations for dissociation reactions

NaCl (s) ---> Na^+ (aq) + Cl^- (aq)

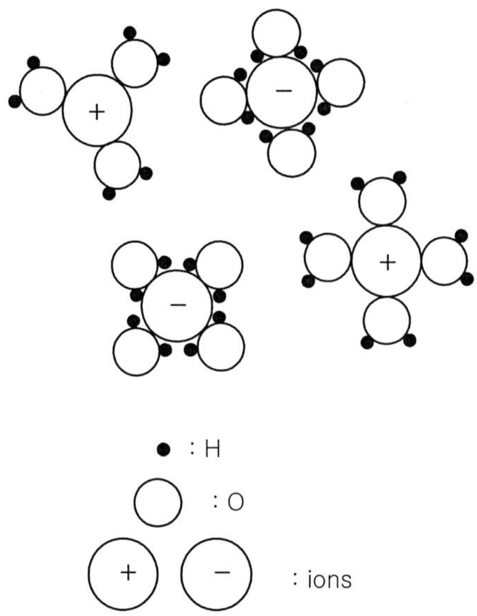

Fig. 5.2 Solvation of ions (hydration) in water.

When a solute particle is surrounded by water molecules, we say it is hydrated(수화).

5.3.2 Equations for ionic reactions

Ex.

molecular equation: $Pb(NO_3)_2$ (aq) + 2KI (aq) ---> PbI_2 (s) + $2KNO_3$ (aq)

precipitate

ionic equation: $Pb^{2+} + 2NO_3^- + 2K^+ + 2I^-$ ---> $PbI_2 + 2K^+ + 2NO_3^-$

○ We can see that the reaction between $Pb(NO_3)_2$ and KI is really a reaction between ions, and can be approximately referred to as an ionic reaction.

○ Ions that do not take part in a reaction (K^+ and NO_3^- in the above example) are sometimes called separator ions; in a sense, they just "stand by and watch the action".

○ net ionic equation: Pb^{2+} (aq) + $2I^-$ (aq) ---> PbI_2 (s)

5.3.3 Acids and bases as electrolytes

In general, acids are molecular substances that react with water to produce ions, one of which is the hydronium ion H_3O^+ (in many times referred to as hydrogen ion H^+).

HCl (g) + H_2O ---> H_3O^- (aq) + Cl^- (aq)

Then the solution conducts electricity and can be called as electrolyte.

○ monoprotic acid; HCl, HNO_3, $HC_2H_3O_2$
polyprotic acid:
diprotic acid; H_2SO_4, H_2CO_3
triprotic acid; H_3PO_4

○ nonmetal oxides as acids.

SO_3 (g) + H_2O ---> H_2SO_4 (aq)
N_2O_5 (g) + H_2O ---> $2HNO_3$ (aq)
CO_2 (g) + H_2O ---> H_2CO_3 (aq)
└-------┘
↳ acidic anhydrides (무수산)

○ A base is a substance that produces hydroxide ion in water, or is able to react with hydronium ion.

○ Solutions of bases contain ions, so bases are also electrolytes.

NaOH (s) ---> Na^+ (aq) + OH^- (aq)
↳ a sort of metal hydroxide

Soluble metal oxides are basic anhydrides (무수염기) because they react with water to form the hydroxide ion as one of the products.

CaO (s) + H_2O ---> $Ca(OH)_2$ (aq)

○ Base + H_2O ---> Base H^+ + OH^-
↳ Ex.; NH_3(g)

5.3.4 Strong and weak acids and bases

○ strong acids (strong electrolyte)
HCl (aq); hydrochloric acid
HNO_3 (aq); nitric acid
H_2SO_4 (aq); sulfuric acid
HBr (aq); hydrobromic acid
HI (aq); hydriodic acid
$HClO_3$ (aq); chloric acid
$HClO_4$ (aq); perchloric acid

○ weak acids (weak electrolyte)
$HC_2H_3O_2$ (aq); acetic acid
H_2CO_3; carbonic acid
HNO_2; nitrous acid

○ strong bases (strong electrolyte)

$NaOH$; sodium hydroxide
KOH; potassium hydroxide
$Ca(OH)_2$; calcium hydroxide
$Ba(OH)_2$; barium hydroxide
$LiOH$; lithium hydroxide
$RbOH$; rubidium hydroxide
$CsOH$; cesium hydroxide
$Sr(OH)_2$; strontium hydroxide

○ weak bases (weak electrolyte)

NH_3 (aq); ammonia

5.3.5 Stoichiometry of ionic reactions

○ 0.1 *M* $CaCl_2$, when dissolved in 1.0L solution, contains 0.1 mol of Ca^+ and 0.2 mol of Cl^-.

$CaCl_2^-$ ---> Ca^+ + $2Cl^-$

5.4 Oxidation-Reduction Reactions

5.4.1 Oxidation-reduction reactions

○ Electron transfer reactions are called oxidation-reduction reactions, or simply *redox* reactions.

oxidation; loss of electrons by one reactant
reduction; gain of electrons by another

Na ---> Na^+ + e^- (oxidation)
Cl_2 + $2e^-$ ---> $2Cl^-$ (reduction)

Ex.; $2Mg$ + O_2 ---> $2MgO$

This can be written as
Mg ---> Mg^{2+} + $2e^-$ (oxidation)
O_2 + $4e^-$ ---> $2O^{2-}$ (reduction)
So, Mg is oxidized (Mg is a reducing agent)
O_2 is reduced (O_2 is an oxidizing agent)

○ Oxidation is an increase in oxidation number.
Reduction is a decrease in oxidation number.

0 0 +1 -1

$H_2 + Cl_2$ ---> 2 H Cl

↳-----------------↗(increase)

↳--------------↗(decrease)

5.4.2 Balancing redox equations by the ion-electron method

○ Basic principle of the ion-electron method

Ex.; reaction of iron(III) chloride $FeCl_3$ with tin(II) chloride $SnCl_2$.

skeleton equation; $Fe^{3+} + Sn^{2+}$ ---> $Fe^{2+} + Sn^{4+}$

└--------┘ └---------┘

↳ reactants ↳ products

half-reaction; Sn^{2+} ---> $Sn^{4+} + 2e^-$

$Fe^{3+} + e^-$ ---> Fe^{2+}

final balanced equation; $2Fe^{3+} + Sn^{2+}$ ---> $2Fe^{2+} + Sn^{4+}$

○ Balancing redox equations for acidic solutions

Ex.; skeleton equation: $CrO_7^{2-} + Fe^{2+}$ ---> $Cr^{3+} + Fe^{3+}$ in acidic solution

final balanced equation: $14H^+ + CrO_7^{2-} + 6Fe^{2+}$ ---> $2Cr^{3+} + 6Fe^{3+} + 7H_2O$

○ Balancing redox equations for basic solutions

Ex.; skeleton equation: $SO_3^{2-} + MnO_4^-$ ---> $SO_4^{2-} + MnO_2$

final equation: $H_2O + 3SO_3^{2-} + 2MnO_4^-$ ---> $3SO_4^{2-} + 2MnO_2 + 2OH^-$

Homework

1. Carbon C forms a compound with an unknown element Y in the formula CY_3. A sample of this compound was found to contain 2g of C and 40g of Y. Calculate the atomic mass of Y. Atomic mass of C is equal to 1.66054×10^{-24} g.

2. In the compound CH_4, the weight percent of all the hydrogens is measured to be 25.15 % relative to the total weight of the compound. The atomic mass of carbon-12 is found to be 1.66054×10^{-24} g. Calculate the atomic mass of the hydrogen.

3. Balance the following equations.
   (a) $Ca(OH)_2 + HCl$ --> $CaCl_2 + H_2O$
   (b) $AgNO_3 + CaCl_2$ --> $Ca(NO_3)_2 + AgCl$
   (c) $C_4H_{10} + O_2$ --> $CO_2 + H_2O$

(d) $CaO + HNO_3$ --> $Ca(NO_3)_2 + H_2O$

(e) $NaHCO_3$ --> $Na_2CO_3 + CO_2 + H_2O$

4. The octane C_8H_{18} in gasoline burns (react with the oxygen O_2) to produce carbon dioxide and steam. (a) Construct the balance equation for the oxidation process. (b) How many moles of the oxygen are needed to react fully with 10 moles of octane?

5. Surfactant SPAN 80, $C_{24}H_{44}O_6$, of molecular mass 428.61 g/mole is to be solved in the dodecane, $C_{12}H_{26}$, of 170.33 g/mole in the weight-percent concentration 5%. We do not have a weight-measurement tool but a volume-measurement tool. (a) For 1 liter of dodecane, how many liters of SPAN 80 should be mixed in order to produce 5% concentration solution? The densities of dodecane and SPAN 80 are found to be 0.750 g/cm^3 and 0.986 g/cm^3, respectively. (b) What is the molarity of the solution?

6. Calculate the number of moles of each of the ions in the following solutions.

(a) 50 mL of 1.2 *M* KOH

(b) 30 mL of 0.5 *M* $CaCl_2$

(c) 10 mL of 0.3 *M* $AlCl_3$

Chapter 6. Physical Chemistry: Solutions of Electrolytes

6.1 Faraday's Law of Electrolysis

The mass of an element produced at an electrode is proportional to the charge (quantity of electricity) Q passed through the liquid; the SI unit of Q is the coulomb (C). The charge is equal to the current I (ampere) multiplied by the time (s):

$$Q = It \tag{6.1}$$

The mass of an element liberated at an electrode is proportional to the equivalent weight of the element and its proportionality factor is known as Faraday constant (F); $F = 96485$ [C/mol]. In other words, 96485 [C] will liberate 1 mol of the element, where the quantity mol is based on the number of elements to be needed (reduction) for the metal to deposit and those to be released (oxidation) for the gas liberation (reduction).

Ex. An aqueous solution of gold(III) nitride, $Au(NO_3)_3$, is electrolyzed with a current of 0.025 [A] until 1.2 [g] of Au ($Ar = 197.0$) is deposited at the cathode (Fig. 6.1). Calculate the charge passed, the duration of the experiment, and the volume of O_2 liberated at the anode at 25°C and 1 [bar] (24.8 [L/mol]).

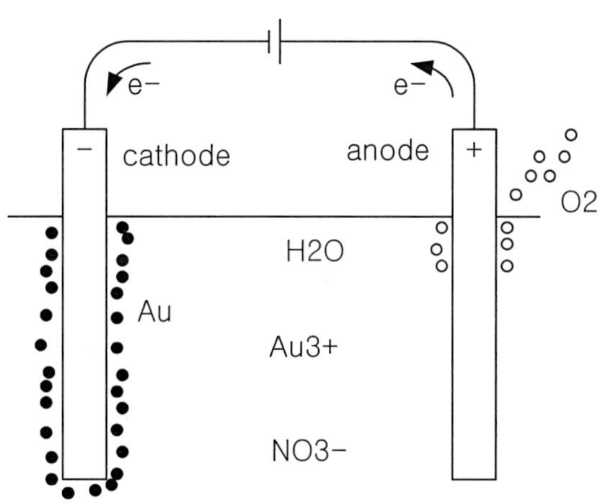

Fig. 6.1 electrolyzation of gold(III) nitride in water.

Reaction at the cathode;

$$3e^- + Au^{3+} \rightarrow Au$$

$$e^- + \frac{1}{3}Au^{3+} \rightarrow \frac{1}{3}Au$$

(96845 C liberates 1/3mol of Au)

Reaction at the anode;

$$\frac{1}{2}H_2O \rightarrow \frac{1}{4}O_2 + H^+ + e^-$$

(96845 [C] liberates 1/4mol of O_2)

The amount of charge needed to produce 1.2 [g] of Au is

$$96845 \times \frac{1.2}{197/3} = 1.76 \times 10^3 \text{ [C]}$$

Time required is thus

$$\frac{1.76 \times 10^3}{0.025} = 7.4 \times 10^4 \text{ [s]} = 20.6 \text{ [h]}$$

From the anode reaction, 96845 [C] liberates 1/4mol of O_2, so that 1.76×10^3 [C] produces

$$\frac{1.76 \times 10^3}{96845 \times 4} = 4.56 \times 10^{-3} \text{ [mol] of } O_2$$

which corresponds to the volume of

$$4.56 \times 10^{-3} \times 24.8 = 0.113 \text{ [L]}$$

6.2 Molar Conductivity

○ electrolyte : a solution of a material forms ions in solution and has a much higher conductivity (e.g. acid etc).

nonelectrolyte : it does not dissociate into ions in a solution and has the same electrical conductivity as water itself (e.g. sucrose, 자당).

○ strong electrolyte : when these are in aqueous solution, they occur almost entirely as ions. e.g. sodium chloride, cupric sulfate.

weak electrolyte : they are present only partially as ions. e.g. acetic acid, ammonia.

○ Ohm's law : $R = V/I$, where R is the resistance in [Ω].

electrical conductance $G = 1/R = I/V$. The conductance is proportional to the area A and inversely proportional to the length L (Fig. 6.2), and here the conductivity κ is defined as follows.

$$G = \kappa A/L \tag{6.2}$$

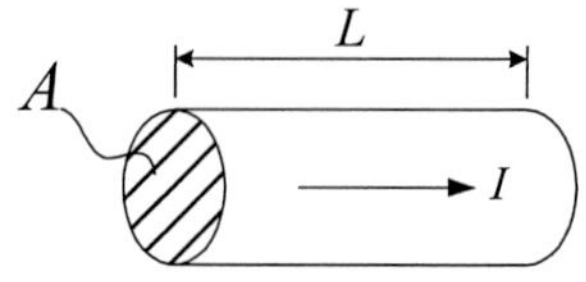

Fig. 6.2 Conduction current through a wire with uniform cross-section.

Note that R or G is dependent not only the nature of the substance but also on the geometry of the substance, while κ is dependent only on the nature of the substance.

○ molar conductivity :

$$\Lambda = \frac{\kappa}{c} \tag{6.3}$$

where c is the concentration of the electrolyte.

Unit of c = [mol/L] or [mol/cc]

Ex.; The electrolyte conductivity κ of a 0.1 M (mol/L) solution of acetic acid was found to be $\kappa = 5.3 \times 10^{-4} [\Omega^{-1}cm^{-1}]$. Calculate the molar conductivity.

$$\Lambda = \frac{\kappa}{c} = \frac{5.3 \times 10^{-4}}{0.1 \times 10^{-3}} = 5.3 [\Omega^{-1}cm^2mol^{-1}]$$

○ Dependence of molar conductivity on the concentration (Fig. 6.3)

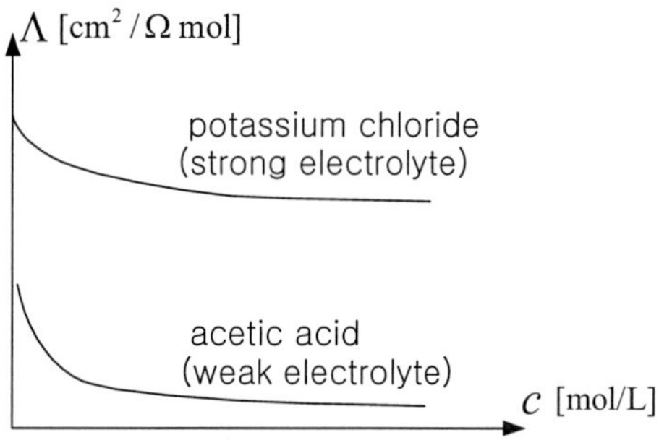

Fig. 6.3 Dependence of molar conductivity on the concentration.

○ Λ_0 : molar conductivity at infinite dilution (i.e. at $c = 0$).

6.3 Weak Electrolyte: The Arrhenius Theory and Ostwald's Law

■ There exists an equilibrium in solution between the undissociated molecules AB and the ions A^+ and B^-;

$$AB \rightleftharpoons A^+ + B^-$$

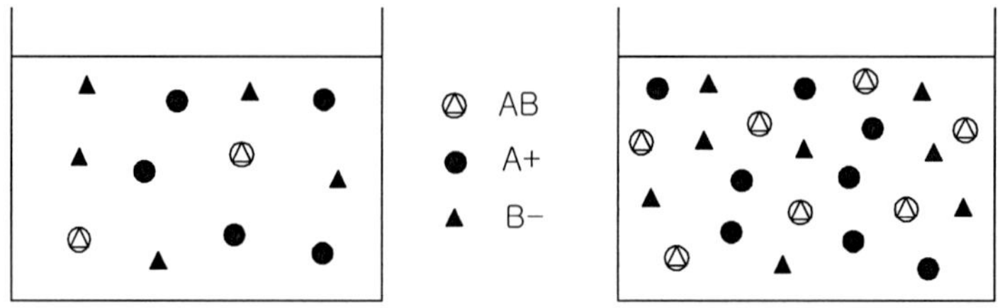

<low concentration>
There are relatively small number of AB in the solution.

<high concentration>
There are relatively large number of AB in the solution.

Fig. 6.4 Equilibrium between undissociated molecules AB and the ions A^+ and B^-.

○ Degree of dissociation, denoted by the symbol α, is defined as the fraction of AB in the form $A^+ + B^-$. Assuming 100% dissociation at infinite dilution, we can write

$$\alpha = \frac{\Lambda}{\Lambda_0} \tag{6.4}$$

At high concentration, there is less dissociation, while at infinite dilution there is complete dissociation and the degree of dissociation is 1 (Fig. 6.4).

○ Ostwald's dilution law

$$AB \rightleftarrows A^+ + B^-$$

The equilibrium constant is defined as

$$K_c = \frac{[A^+][B^-]}{[AB]} \tag{6.5}$$

where [ ] indicates the concentration of each substance in [mol/dm^3] or [mol/cc] etc. Suppose that an amount of n [mol] of the electrolyte is present in a volume V and the fraction dissociated is α.

	AB	$\rightleftarrows$	A^+	+	B^-
amounts present at equilibrium	$n(1-\alpha)$		$n\alpha$		$n\alpha$
concentration at equilibrium	$\frac{n(1-\alpha)}{V}$		$\frac{n\alpha}{V}$		$\frac{n\alpha}{V}$

But $n/V = c$. Therefore

$$\frac{\alpha^2}{1-\alpha} = \frac{K}{c} \tag{6.6}$$

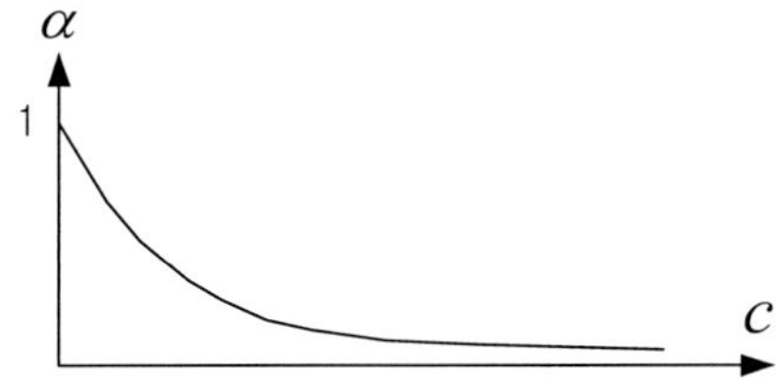

Fig. 6.5 Degree of dissociation as a function of concentration.

Homework

Derive the relation between α and c and show the qualitative dependence of α on the concentration c for a constant parameter K as shown in Fig. 6.5.

6.4 Strong Electrolytes

○ The fall in Λ with increasing concentration must, for strong electrolytes, be attributed to some cause other than a decrease in the degree of dissociation, because strong electrolyte dissociates completely at any concentration.

○ Debye-Hückel Theory

The decrease in the molar conductivity Λ of a strong electrolyte is attributed to the mutual interference of ions, which becomes more pronounced as the concentration increases.

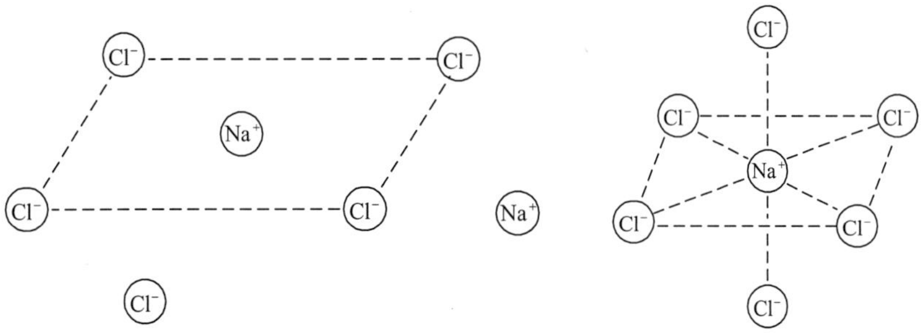

Distribution of Cl^- around a Na^+ for NaCl solution

Distribution of Cl^- around a Na^+ for NaCl solid

Fig. 6.6 Difference in the distribution pattern of ions in the solution (left) and as the solid (right) state.

In solution, the immediate neighborhood of any positive ion tends to have more negative than positive ions, whereas that of a negative ion tends to have more positive than negative ions. If an electric potential is applied, a positive ion moves toward the negative electrode, and vice versa, and its speed is directly related to the conductivity. There are two effects which are related to the ion's motion.

○ Relaxation (or symmetry) effect:

After the + ion moved under an external field to the right-hand side, the ionic atmosphere governed by - ions exerts a pulling force to the + ion (Fig. 6.7). This influence on the speed of the + ion is called the relaxation or asymmetric effect.

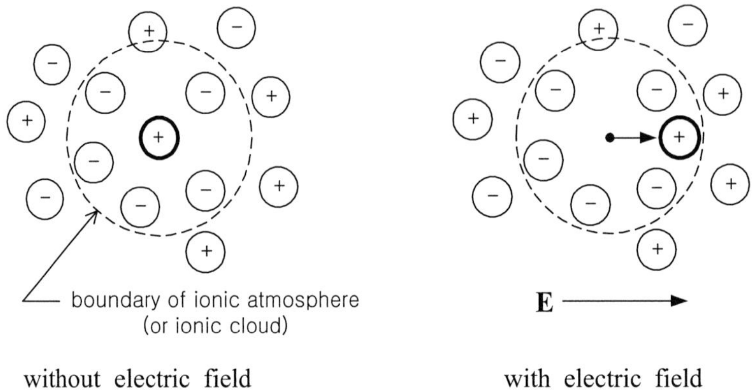

Fig. 6.7 Asymmetric (relaxation) effect in the presence of the external field.

○ Electrophoretic effect.

When the electric potential is applied, the atmosphere must also move to the other direction and exerts viscous-like force to the + ion retarding the + ion's motion.

○ The ionic atmosphere.

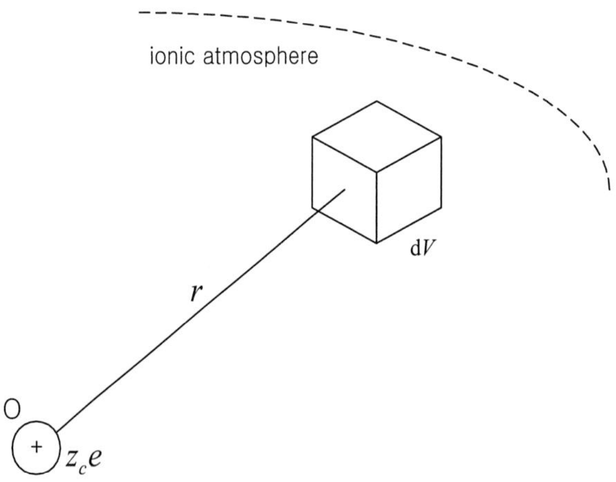

Fig. 6.8 Ionic atmosphere around a positive ion at the point O.

Consider a positive ion at the point O (Fig. 6.8) with the charge $z_c e$, where z_c is the valence of the ion and e is the unit positive (electron) charge. Within the atmosphere, centered at O, we also consider an infinitesimal volume dV. The probability that there exists a negative ion in the volume is greater than the probability that there exists a positive ion. The work required to bring a positive ion of charge z_+e from infinity up to this volume element is $z_+e\phi$ where z_+ is the ionic valence and ϕ is the average electric potential in the volume element. Similarly the work required to bring a negative ion becomes $-z_-e\phi$ where z_- is the absolute value of the valence of the negative ion. The time-average numbers of the positive ions and negative ions in dV are given by the Boltzmann principle (or Boltzmann distribution law):

$$dN_+ = N_+ e^{-z_+e\phi/k_BT} dV$$

$$dN_- = N_- e^{+z_-e\phi/k_BT} dV$$

where N_+ and N_- are the total numbers of positive and negative ions, respectively, per unit volume of solution, k_B is the Boltzmann constant, and T is the absolute temperature. The charge density ρ (i.e. the net charge per unit volume) is

$$\rho = e\frac{(z_+dN_+ - z_-dN_-)}{dV} = e\left(N_+z_+e^{-z_+e\phi/k_BT} - N_-z_-e^{+z_-e\phi/k_BT}\right) \qquad (6.7)$$

If $z_+e\phi/k_BT$ and $z_-e\phi/k_BT$ are small enough (or in other words the ions are not too close to the ion at P), then

$$e^{-z_+e\phi/k_BT} \cong 1 - z_+e\phi/k_BT, \qquad e^{+z_-e\phi/k_BT} \cong 1 + z_-e\phi/k_BT$$

Therefore

$$\rho \cong (N_+z_+ - N_-z_-)e - \frac{e^2\phi}{k_BT}(N_+z_+ + N_-z_-)$$

Conservation of the electric charge requires $N_+z_+ - N_-z_- = 0$. Thus,

$$\rho = -\frac{e^2\phi}{k_BT}(N_+z_+ + N_-z_-)$$

When there are many different types of ions,

$$\rho = -\frac{e^2\phi}{k_BT}\sum_i N_iz_i \qquad (6.8)$$

On the other hand, the Poisson equation which relates ϕ with ρ is

$$\nabla^2\phi = -\frac{\rho}{\epsilon_0\epsilon} \qquad (6.9)$$

In the spherical coordinates with full symmetric properties, this becomes

$$\frac{1}{r^2}\frac{\partial}{\partial r}\left(r^2\frac{\partial\phi}{\partial r}\right) = \frac{-\rho}{\epsilon_0\epsilon}$$

Now, substitute the equation for ρ to obtain

$$\frac{1}{r^2}\frac{\partial}{\partial r}\left(r^2\frac{\partial \phi}{\partial r}\right)=\kappa^2\phi \tag{6.10}$$

where

$$\kappa^2=\frac{e^2}{\epsilon_0\epsilon k_B T}\sum_i N_i z_i^2 \tag{6.11}$$

The general solution is

$$\phi=\frac{Ae^{-\kappa r}}{r}+\frac{Be^{+\kappa r}}{r}$$

We put $B=0$ (because at infinity ρ should be zero);

$$\phi=\frac{Ae^{-\kappa r}}{r}$$

It is clear that as $r\to 0$, the Coulomb's law must be recovered. Thus,

$$\phi=\frac{z_c e\, e^{-\kappa r}}{4\pi\epsilon_0\epsilon r} \tag{6.12}$$

Therefore the factor $e^{-\kappa r}$ reflects the effect of the ionic atmosphere. Here $1/\kappa$ is called 'thickness of the ionic atmosphere (or cloud)' ;

$$1/\kappa=\left(\frac{\epsilon_0\epsilon k_B T}{e^2\sum_i c_i z_i^2 L}\right)^{1/2} \tag{6.13}$$

where $c_i=N_i/L$ is the ion's molar concentration and L is the Avogadro constant, $L=6.022\times10^{23}$ [mol^{-1}].

Exercise

Estimate the thickness of the ionic atmosphere for a solution of

(a) 0.01 ***M*** NaCl and

(b) 0.001 ***M*** $ZnCl_2$, both in water at 25°C with $\epsilon=78$. Use $e=1.602\times10^{-19}$ [C], $k_B=1.381\times10^{-23}$ [J/K], and $\epsilon_0=8.854\times10^{-12}$ [C^2/Nm^2]

Solution

(a) $c_+=0.01$ ***M***, $c_-=0.01$ ***M***

$z_+=1$ (Na^+), $z_-=1$ (Cl^-)

$\sum_i c_i z_i^2=0.01\times1^2+0.01\times1^2=0.02$ [mol/dm^3]

$$\frac{1}{\kappa}=\left[\frac{8.854\times10^{-12}\times78\times1.381\times10^{-23}\times298}{(1.602\times10^{-19})^2\times0.02\times1000\times6.022\times10^{23}}\right]^{1/2}$$

$$=3.03\times10^{-9}[\text{m}]=3.03[\text{nm}]$$

(b) $c_+=0.001$ ***M***, $c_-=0.002$ ***M***

$z_+=2$ (Zn^{++}), $z_-=1$ (Cl^-)

$$\sum_i c_i z_i^2 = 0.001\times 2^2 + 0.002\times 1^2 = 0.006 \ \ [\mathrm{mol/dm^3}]$$

$$\frac{1}{\kappa} = \left[\frac{8.854\times 10^{-12}\times 78\times 1.381\times 10^{-23}\times 298}{(1.602\times 10^{-19})^2\times 0.006\times 1000\times 6.022\times 10^{23}}\right]^{1/2}$$

$$= 5.53\times 10^{-9}[\mathrm{m}] = 5.53[\mathrm{nm}]$$

○ Conductivity of strong electrolytes.

Debye-Hückel-Onsager equation for Λ;

$$\Lambda = \Lambda_0 - (P + Q\Lambda_0)\sqrt{c} \tag{6.14}$$

where c is the concentration, and for a symmetrical electrolyte (i.e. for $z_+ = z_- = z$),

$$P = \frac{zeF}{3\pi\eta}\left(\frac{2z^2e^2L}{\epsilon_0\epsilon k_BT}\right)^{1/2} \tag{6.15}$$

$$Q = \frac{z^2e^2w}{24\pi\epsilon_0\epsilon k_BT}\left(\frac{2z^2e^2L}{\pi\epsilon_0\epsilon k_BT}\right)^{1/2} \tag{6.16}$$

where η is the viscosity of the medium, w is a constant depending on the type of electrolyte (for a uni-univalent electrolyte, $w = 2-\sqrt{2}$), and F is the Faraday constant, $F = 96485$ [C/mol]. This equation is found to be accurate for dilute solutions.

○ Thus, for strong electrolyte, the molar conductivity Λ decreases with c not because of decreased dissociation (dissociation is 100% for any c) but because of decreased mobility.

Homework

Solve the 1-D Poisson equation in the spherical coordinates and derive the formula for the potential $\phi(r)$ and the thickness of the ionic atmosphere $1/\kappa$ for the strong electrolyte.

6.5 Independent Migration of Ions

○ Kohlrausch's law of independent migration of ions:

Each ion is assumed to make its own contribution to the molar conductivity, irrespective of the nature of the other ion with which it is associated.

$$\Lambda_0 = \lambda_+^0 + \lambda_-^0 \tag{6.17}$$

where λ_+^0 and λ_-^0 are the ion conductivities of cation and anion, respectively, at infinite dilution.

Table 1. Molar conductivity of typical electrolytes.

Electrolyte	Λ_0 [Ωcm²/mol]	Electrolyte	Λ_0 [Ωcm²/mol]	difference
KCl	149.8	NaCl	126.4	23.4
KI	150.3	NaI	126.9	23.4
$\frac{1}{2}K_2SO_4$	153.5	$\frac{1}{2}Na_2SO_4$	130.1	23.4

In the above table, we can see invariant differences even for different anions. This means that each ion must have independent effect on the conductivity. For example

$$\Lambda_0(\text{KCl}) = \lambda_0(\text{K}^+) + \lambda_0(\text{Cl}^-) = 149.8$$

$$\Lambda_0(\text{NaCl}) = \lambda_0(\text{Na}^+) + \lambda_0(\text{Cl}^-) = 126.4$$

$$\therefore\ \lambda_0(\text{K}^+) - \lambda_0(\text{Na}^+) = 23.4 \quad [\Omega\text{cm}^2/\text{mol}]$$

This difference is invariant of the nature of the anion. That means $\lambda_0(K^+)$ and $\lambda_0(Na^+)$ each has its own characteristic value as shown in Table 2.

Table 2. Limiting molar conductivity of typical ions

cation	λ_0^+ [Ωcm²/mol]	anion	λ_0^- [Ωcm²/mol]
H^+	349.8	OH^-	197.6
Na^+	50.1	Cl^-	76.3
K^+	73.5	NO_3^-	71.4
Ag^+	61.9	$\frac{1}{2}SO_4^{2-}$	80.0
$\frac{1}{2}Mg^{2+}$	53.1	CH_3COO^-	40.9
$\frac{1}{2}Ca^{2+}$	59.5	I^-	76.8

○ Ionic mobilities.

u (ion mobility) = speed with which the ion moves under a unit electric field. Assume that there exist univalent positive and negative ions in a unit cube (Fig. 6.9). The potential difference between two facing planes is denoted as V.

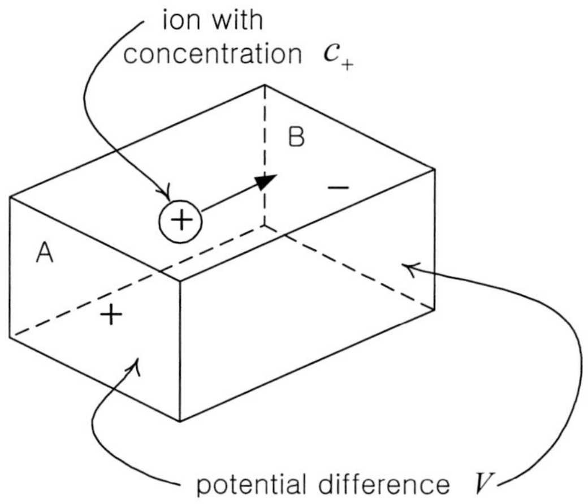

Fig. 6.9 Effect of the potential difference V to the ion's motion in a unit cube.

Then, the speed of the ions is $u_+ V$. Number of positive ions that pass the plane A (or B) during a unit time is $u_+ V c_+$ where c_+ is the ion concentration (number of ions per unit volume). The charge they carry is $F u_+ V c_+$, so that the current is also $F u_+ V c_+$. The electrolyte conductivity κ_+ due to the positive ions is thus

$$\kappa_+ = \frac{\text{current}}{\text{potential diff.}} = \frac{F u_+ V c_+}{V} = F u_+ c_+ \tag{6.18}$$

The molar conductivity due to these positive ions is

$$\lambda_+^0 = \frac{\kappa_+}{c_+} = F u_+ \tag{6.19}$$

○ Ion mobilities and diffusion coefficient

Ions migrate in the absence of an electric field, if there is a concentration gradient. Fick's law of diffusion for the concentration c is

$$\frac{\partial c}{\partial t} = \nabla \cdot (D \nabla c) \tag{6.20}$$

where D is the diffusion coefficient. On the other hand, Nernst showed that D is given by

$$D = \frac{k_B T}{Q} u \tag{6.21}$$

where u is the mobility of an ion, T is the temperature and Q is the charge on the ion. Further we use the following relation

$$Q = F \mid z \mid / L \tag{6.22}$$

to obtain

$$D = \frac{k_B T \lambda L}{F^2 \mid z \mid} = \frac{R T \lambda}{F^2 \mid z \mid} \tag{6.23}$$

where R is the gas constant; $k_B = R/L$. These relationships show that if we know the diffusivity and conductivity through experiments, we can calculate the mobility. For the uni-univalent electrolyte such as NaCl, the overall diffusion coefficient D can be obtained from

$$D = \frac{D_A^+ D_B^-}{D_A^+ + D_B^-} \tag{6.24}$$

where D_A^+ and D_B^+ indicate the diffusion coefficients of separate ions A^+ and B^-.

Homework

Derive the formula relating the diffusion coefficient D and the molar conductivity λ shown above.

6.6 Activity Coefficient

○ Ionic strength.

$$I = \frac{1}{2}\sum_i c_i z_i^2 \tag{6.25}$$

ex 1) uni-univalent electrolyte (e.g. NaCl);

$z_+ = 1,\ z_- = 1,\ c_+ = c_- = c$

$$I = \frac{1}{2}(c_+ z_+^2 + c_- z_-^2) = c$$

ex 2) uni-bivalent electrolyte (e.g. K_2SO_4);

$z_+ = 1,\ z_- = 2,\ c_+ = 2c,\ c_- = c$

$$I = \frac{1}{2}(2c \times 1^2 + c \times 2^2) = 3c$$

ex 3) uni-trivalent electrolyte (e.g. Na_3PO_4);

$z_+ = 1,\ z_- = 3,\ c_+ = 3c,\ c_- = c$

$$I = \frac{1}{2}(3c \times 1^2 + c \times 3^2) = 6c$$

Chapter 7. Electrostatics

7.1 Electric Charge

○ Experiment: rub a glass rod with silk and lift (isolate) it (Fig. 7.1). Then the glass rod obtains tiny amount of positive charge from the silk.

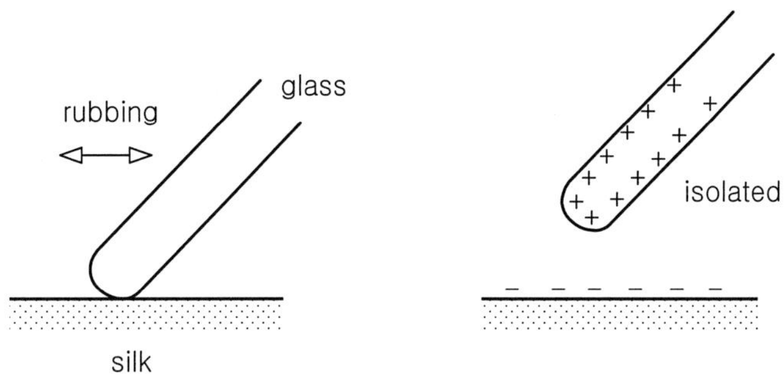

Fig. 7.1 Rubbing of a glass rod on silk cloth (left) and abrupt isolation of the rod from the silk for charging.

On the other hand, rub a plastic rod with a fur, then the plastic rod is charged negative. Bring the plastic rod near the glass rod, then the two rods attract each other.

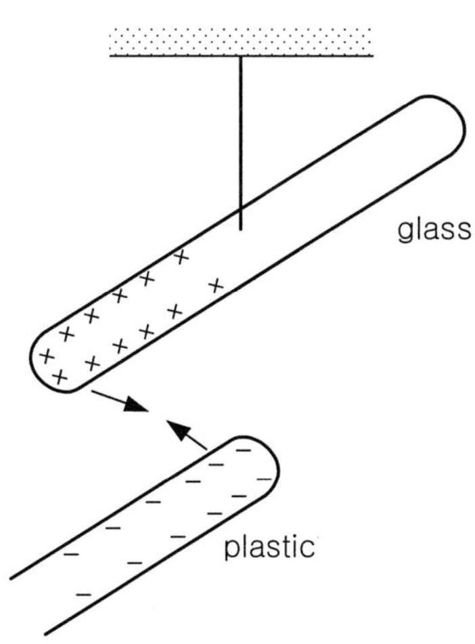

Fig. 7.2 Attracting electrostatic force between positively and negatively charged rods.

This force is called "electrostatic" force. [Here "static" means that the charges are either stationary or moving very slowly.]

7.2 Conductor and Insulator

○ Conductor: the material in which some of the negative charge can move rather freely; metals, tab water, human body etc.

○ Insulator (Nonconductor): the material in which none of the negative charges can move freely.

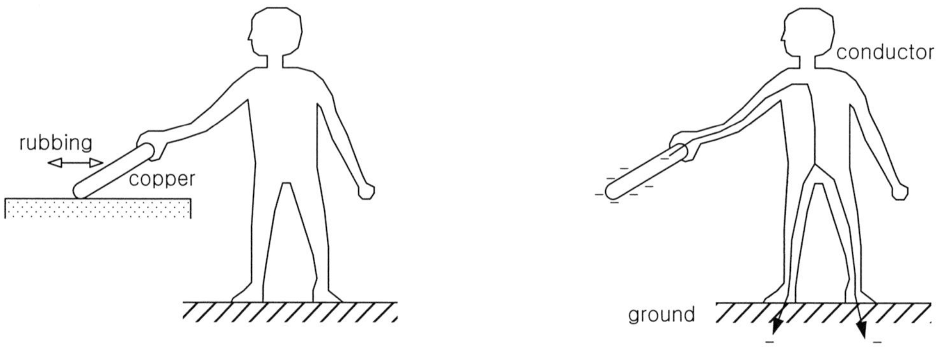

Fig. 7.3 Human body can be considered to be a kind of conductor.

When the charged plastic rod is brought near the copper rod which was in a neutral state, then each end of the copper rod is charged with opposite sign (Fig. 7.4).

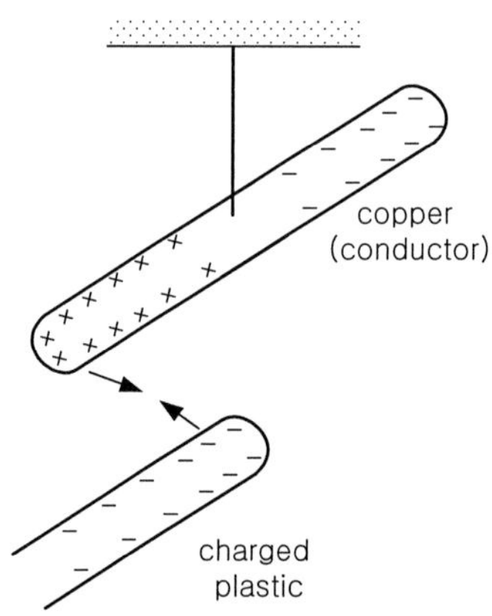

Fig. 7.4 Positive and negative charges are induced in the close and far ends of the coper rod, respectively, when the negatively charged plastic rod is brought.

If the plastic rod is brought near the copper rod, the conduction electrons in the copper rod's closer end are repelled by the negative charge on the plastic rod, leaving the atoms in that region charged positively.

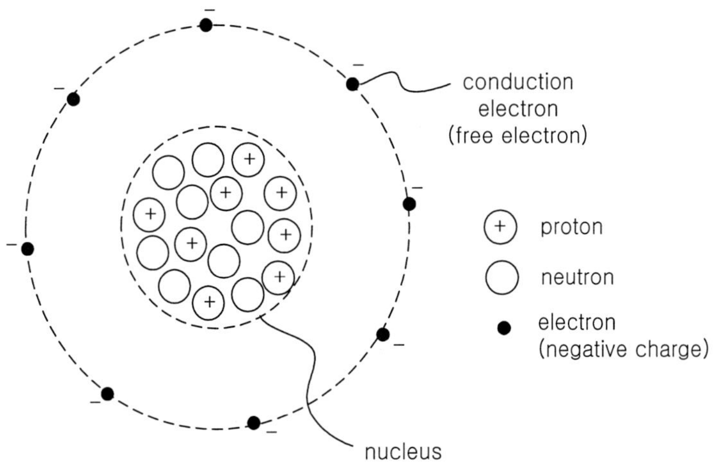

Fig. 7.5 Free electrons located at the outermost electron orbit.

After the conduction electrons are moved, the left-hand side is charged positively and the right-hand side negatively (Fig. 7.6).

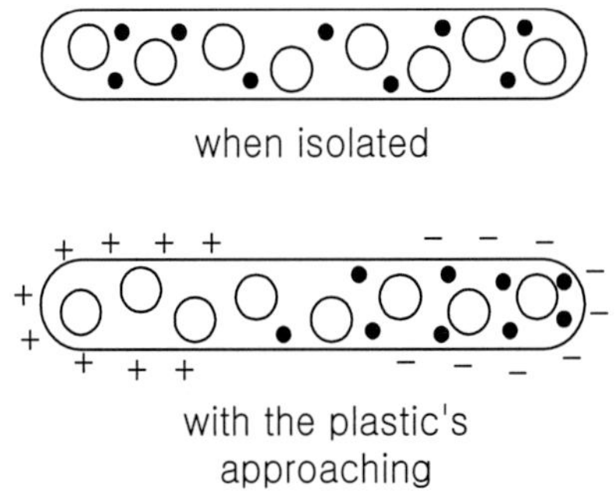

Fig. 7.6 Initially uniformly distributed free electrons in the copper (top) and asymmetrically distributed free electrons when the plastic rod approaches the copper (bottom).

○ Semiconductor: material that is electrically intermediate between conductor and insulator, e.g., silicon and germanium.

○ Superconductor: material that presents no resistance to the movement of electric charge.

○ Electric current: the phenomenon of the electrons' motion through the materials (opposite way).

7.3 Coulomb's Law

○ Unit of charge: C (coulomb)

One coulomb is the amount of charge that is transferred through the cross-section of a wire in 1 second, where there is an electric current of 1 ampere in the wire.

$$i = \frac{dq}{dt} \tag{7.1}$$

○ Analogy between the electricity and fluid flow.

electricity		fluid flow
current (i)	<--->	flow rate (Q)
charge (q)	<--->	volume (V)
time (t)	<--->	time (t)

○ Coulomb's law:

$$F = k\frac{|q_1||q_2|}{r^2} \tag{7.2}$$

k; electrostatic constant

q_1, q_2; charge

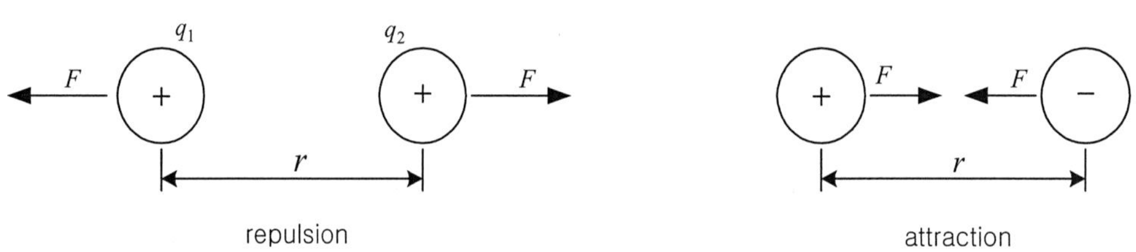

Fig. 7.7 Coulomb forces repulsing when both charges have the same sign and attracting when both have different sign of polarity.

$$k = \frac{1}{4\pi\epsilon_0} = 8.99 \times 10^9 \ \mathrm{Nm^2/C^2}$$

ϵ_0 (permittivity of vacuum) $= 8.85 \times 10^{-12} \ \mathrm{C^2/Nm^2}$

○ Charge is quantized:

$$q = ne \tag{7.3}$$

Here n is an integer and e is the elementary charge;

$$e = 1.60 \times 10^{-19} \ \mathrm{C}$$

○ Charge is conserved.

7.4 Electric Fields

○ Electric field $\boldsymbol{E}$: the electrostatic force vector $\boldsymbol{F}$ that acts on the test charge of 1 coulomb. The unit is N/C or V/m.

$$\boldsymbol{E} = \frac{\boldsymbol{F}}{q_0} \qquad (q_0 = 1 \text{ coulomb}) \tag{7.4}$$

For a point charge we have

$$\boldsymbol{E} = \frac{q}{4\pi\epsilon_0 r^2}\hat{\boldsymbol{r}} \tag{7.5}$$

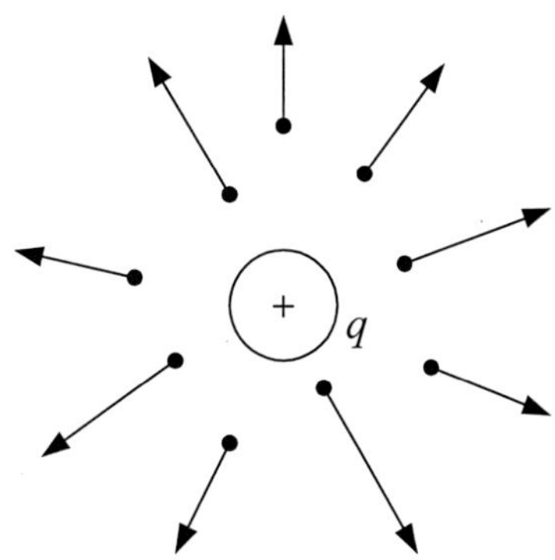

Fig. 7.8 Electric field vectors at various points around a positively charged particle with q.

○ Electric field lines: lines connecting the electric field vectors $\boldsymbol{E}$

Electric field lines extend away from positive charge (Fig. 7.9) and run toward negative charge.

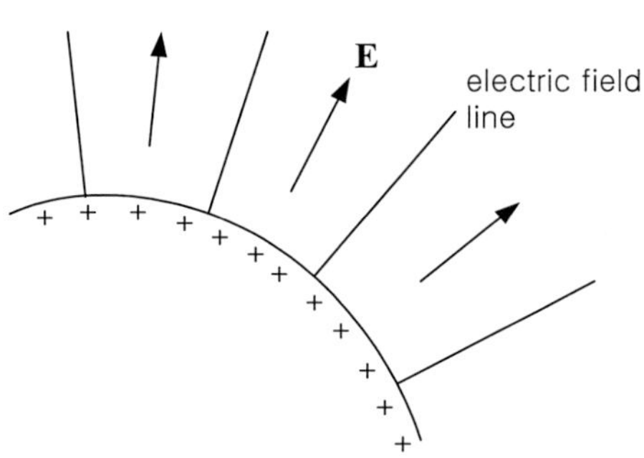

Fig. 7.9 Electric field vectors and lines near a positively charged surface.

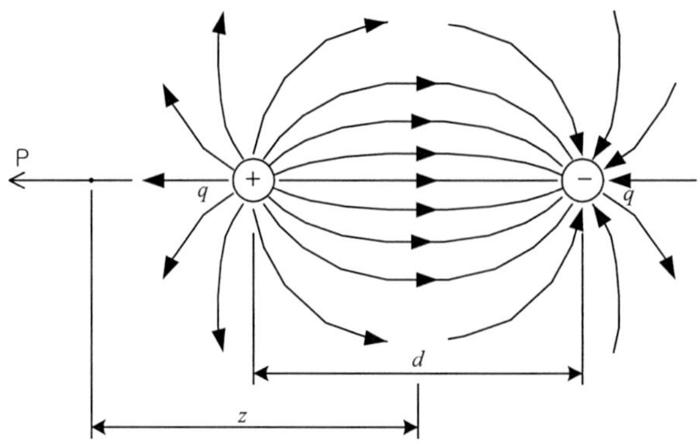

Fig. 7.10 Electric field lines near an electric dipole

It can be shown that far from the dipole, E at a certain point at the axis P (Fig. 7.10) is given by

$$E = \frac{1}{2\pi\epsilon_0}\frac{p}{z^3} \tag{7.6}$$

where p is the magnitude of the electric dipole moment $\boldsymbol{p} = qd\boldsymbol{i}$.

○ Electrostatic force $\boldsymbol{F}$ experienced by a point charge q in an electric field $\boldsymbol{E}$ is given by

$$\boldsymbol{F} = q\boldsymbol{E} \tag{7.7}$$

○ Ink-jet printing

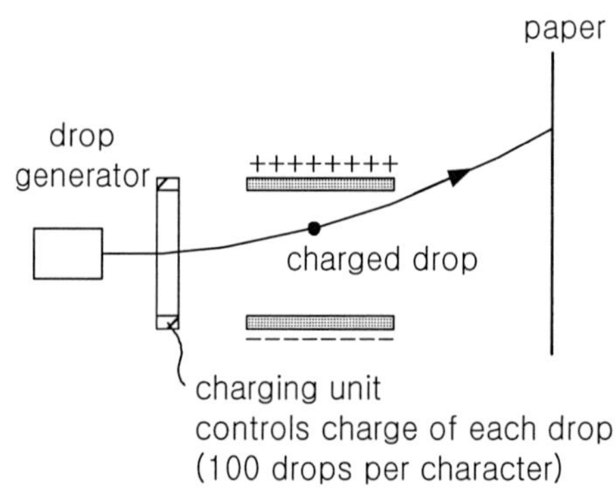

Fig. 7.11 Principle of ink-jet printing.

○ Some measures of electric charge

symbol (name)	SI unit
q (charge)	C
λ (linear charge density)	C/m
σ (surface charge density)	C/m^2
ρ (volume charge density)	C/m^3

○ Permanent dipole in an electric field

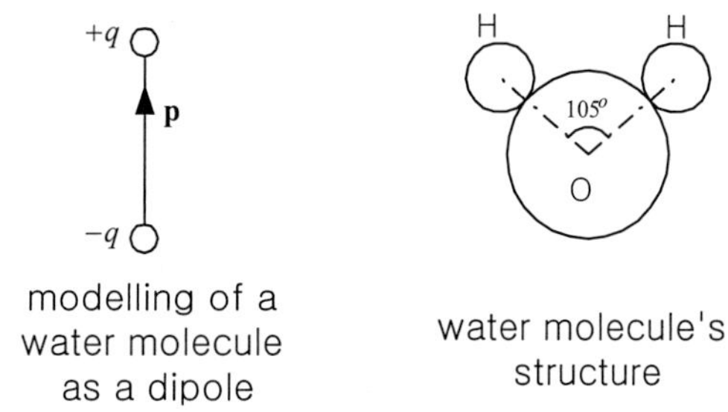

Fig. 7.12 Structure of a water molecule (right) and its modelling with a dipole (left).

$\boldsymbol{F} = q\boldsymbol{E}$

torque $\tau = Fd\sin\theta$

But $F = qE$

$p = qd$

$\therefore\ \tau = pE\sin\theta$

$\boldsymbol{\tau} = \boldsymbol{p} \times \boldsymbol{E}$ (torque vector on a dipole)

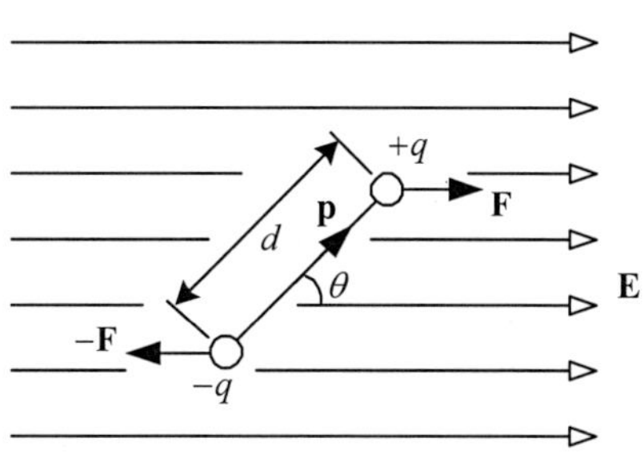

Fig. 7.13 Schematic illustrating how torque is generated for a dipole when subjected to an external field.

Potential energy of a dipole in an electric field $\boldsymbol{E}$ corresponds to the work necessary to turn the dipole from 90° $(=\theta_0)$ to an arbitrary angle θ (Fig. 7.13).

$$U = -W = -\int_{\pi/2}^{\theta} \tau d\theta = \int_{\pi/2}^{\theta} pE\sin\theta\, d\theta = -pE\cos\theta = \boldsymbol{p} \cdot \boldsymbol{E} \qquad (7.8)$$

○ Microwave oven: oscillating electric field causes electric dipoles to vibrate. The microwaves in home ovens are tuned to the natural frequency of water molecules' vibration.

7.5 Gauss' Law

○ Flux of an electric field:

$$\Phi_{\text{net}} = \oint \boldsymbol{E} \cdot \mathrm{dA} \tag{7.9}$$

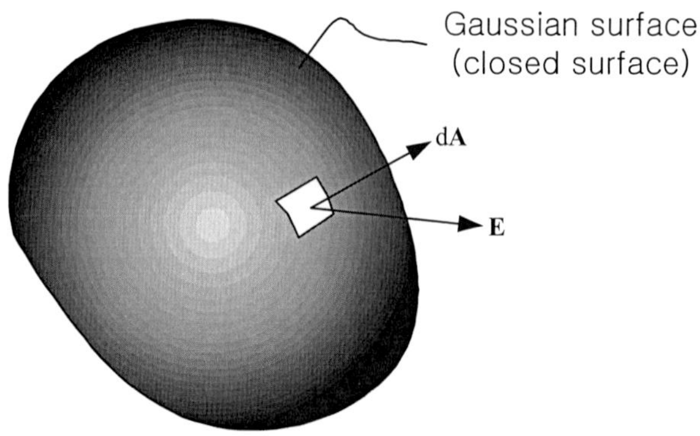

Fig. 7.14 Gaussian surface and its small element for calculation of electric-field flux using Eq. (7.9).

○ Gauss' law:

Net charge q_{enc} within the Gaussian surface is given by

$$q_{\text{enc}} = \epsilon_0 \Phi_{\text{net}} = \epsilon_0 \oint \mathrm{E} \cdot \mathrm{dA} \tag{7.10}$$

This equation however holds only when the net charge is located in a vacuum or in air.

$$\oint_{S1} \boldsymbol{E} \cdot \mathrm{dA} > 0$$

$$\oint_{S2} \boldsymbol{E} \cdot \mathrm{dA} < 0$$

$$\oint_{S3} \boldsymbol{E} \cdot \mathrm{dA} = \oint_{S4} \mathrm{E} \cdot \mathrm{dA} = 0$$

Gauss' law and Coulomb's law are equivalent !

Example

In Fig. 7.15:

Through S1; field is totally coming out --> charge inside S1 is positive

Through S2; field is totally coming in --> charge inside S1 is negative

Through S3; field is partially coming out and coming in leading to zero total flux --> no charge inside S3.

Through S4; if sum of positive and negative charges is zero, total flux must be zero.

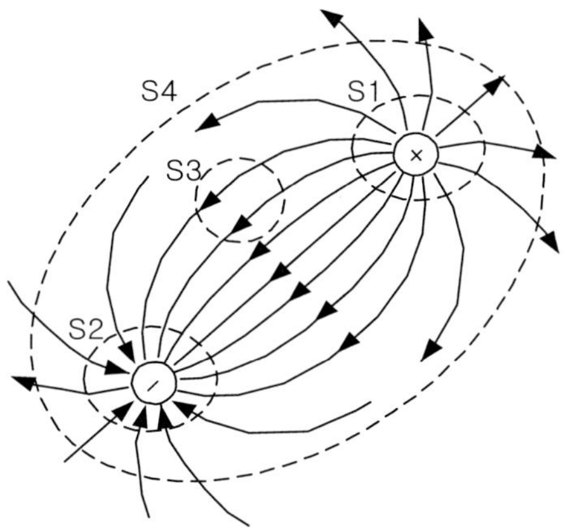

Fig. 7.15 Various kinds of Gaussian surfaces where Gauss' law applies.

○ An isolated charged conductor

If an excess charge is placed in an isolated conductor, that amount of charge will move entirely to the surface of the conductor (Fig. 7.16). None of the excess charge will be found within the body of the conductor. The electric field must be normal to the surface.

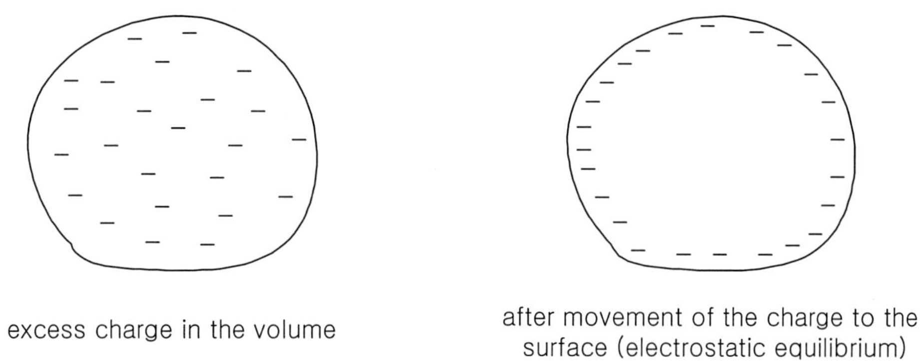

Fig. 7.16 Distribution of negative charges in a conductor: unrealistic instantaneous volumetric distribution inside a conductor (left) and realistic distribution on the surface.

Gauss' law: $\Delta q = \epsilon_0 \int \boldsymbol{E} \cdot \mathrm{d}\boldsymbol{A} = \epsilon_0 E \Delta A$

On the other hand, $\Delta q = \sigma \Delta A$

$$\therefore \; E = \frac{\sigma}{\epsilon_0} \tag{7.11}$$

where σ is the surface charge density (Fig. 7.17).

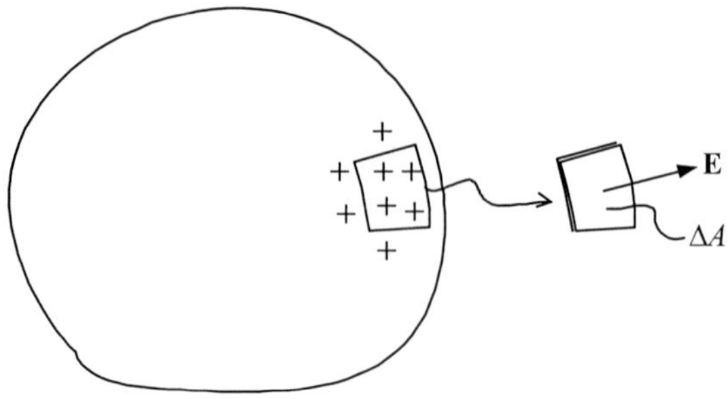

Fig. 7.17 Schematic showing a small surface element with surface-charge distribution creating the field $\boldsymbol{E}$.

※ Derivation of Gauss' law for a point charge

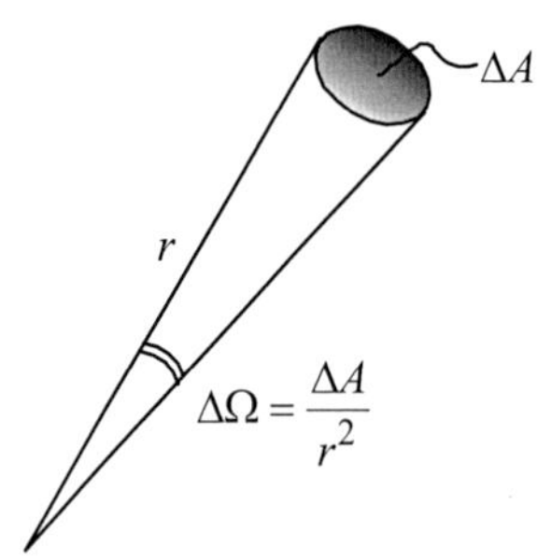

Fig. 7.18 Definition of solid angle $\Delta\Omega$

For the entire spherical surface;

$$\Omega = \int \frac{\mathrm{dA}}{r^2} = \frac{1}{r^2} 4\pi r^2 = 4\pi \ \text{[steradian]}$$

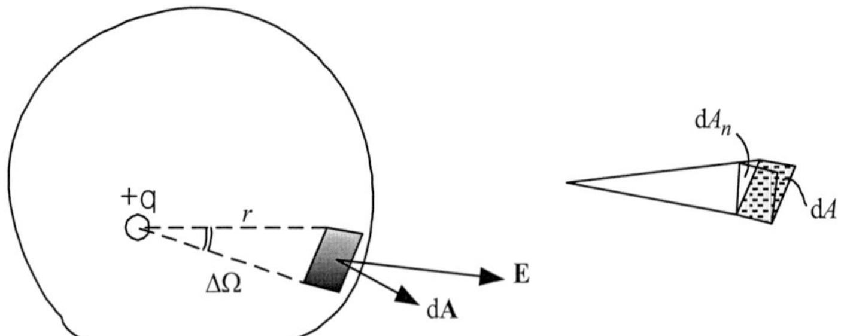

Fig. 7.19 Small surface element with the solid angle $\Delta\Omega$ to be used in evaluation of the field flux Φ_{net}.

$$\Delta\Phi = \boldsymbol{E} \cdot d\boldsymbol{A} = k\frac{q}{r^2}\hat{\boldsymbol{r}} \cdot d\boldsymbol{A}$$

$$= k\frac{q}{r^2}dA_n = kq\Delta\Omega$$

$$\Phi_{\text{net}} = \oint \boldsymbol{E}\cdot d\boldsymbol{A} = \oint kq\,\mathrm{d}\Omega = 4\pi kq = \frac{q}{\epsilon_0}$$

where $4\pi k = 1/\epsilon_0$ is used and $\hat{\boldsymbol{r}}$ denotes the unit vector along the r-direction.

○ Electric field around a linear charge (Fig. 7.20)

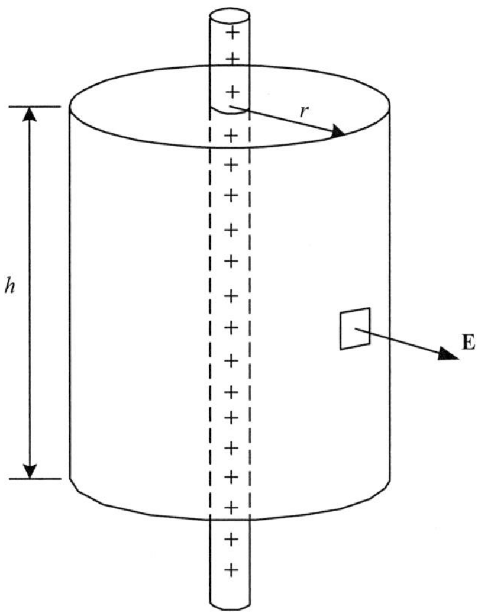

Fig. 7.20 Gauss surface for calculating the field around a linear charge.

$$\Phi = E(2\pi rh)$$

$$\Phi = \frac{q_{\text{net}}}{\epsilon_0} = \frac{\lambda h}{\epsilon_0} \quad \text{(Gauss' law)}$$

$$\therefore\ E(2\pi rh) = \frac{\lambda h}{\epsilon_0}$$

$$E = \frac{\lambda}{2\pi\epsilon_0 r} \tag{7.12}$$

where λ is the linear charge density.

7.6 Electric Potential

○ Potential energy

Let W_∞ represent the work done by the electrostatic force acting on a charged particle q_0 (Fig. 7.21). Then the electric potential energy U is given by

$$U = -W_\infty = -\int_\infty^r q_0 \boldsymbol{E} \cdot d\boldsymbol{r} \tag{7.13}$$

which is path-independent (Coulomb force is conservative). For a given point charge q, we have $E = kq/r^2$, and $\boldsymbol{E} \cdot d\boldsymbol{r}$ can be replaced by $E\,dr$ so that

$$U = -\int_\infty^r q_0 q k \frac{dr}{r^2} = \left[\frac{kq_0 q}{r}\right]_\infty^r = \frac{kq_0 q}{r} \tag{7.14}$$

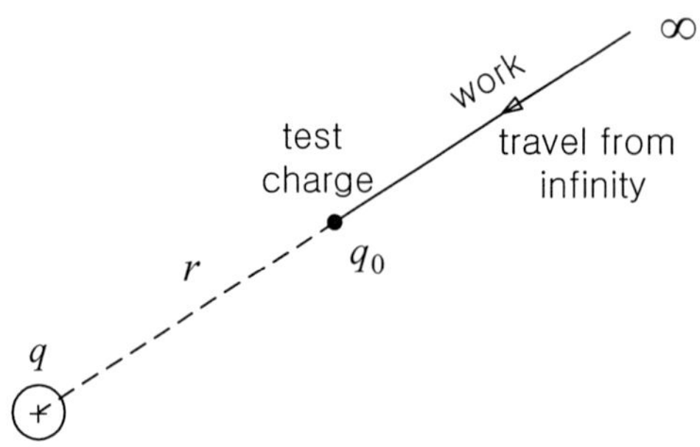

Fig. 7.21 Work done by a test charge during the travel from infinity.

○ Electric potential

$$V = \frac{U}{q_0} \tag{7.15}$$

For a point charge; $V = kq/r$ (independent of the test charge)

The electric potential can be understood as electric potential energy for a unit charge.

Unit; volt = J/C = Nm/C

1 electron-volt (eV) = $e \times 1\text{V} = 1.60 \times 10^{-19}\text{C} \times 1\text{J/C} = 1.60 \times 10^{-19}\text{J}$ (energy)

○ Work done by an applied force (Fig. 7.22)

$$W_{\text{app}} = U_f - U_i = q\Delta V \tag{7.16}$$

where $\Delta V = V_f - V_i$.

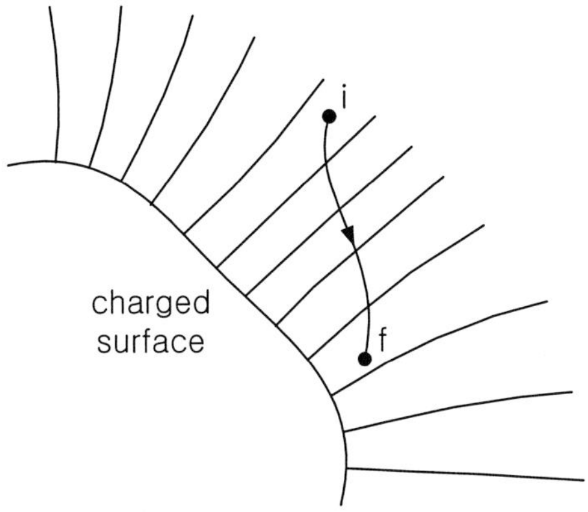

Fig. 7.22 An arbitrary path of a charge's movement in the field.

○ Equi-potential surfaces and electric field lines (Fig. 7.23)

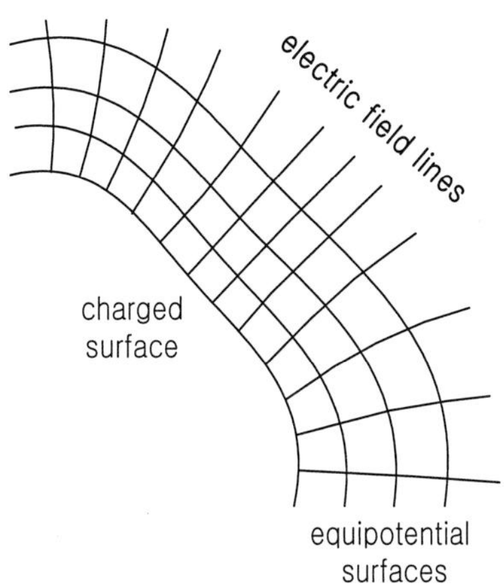

Fig. 7.23 Equi-potential surfaces and the electric field lines.

○ Induced dipole moment (Fig. 7.24): Nonpolar molecules have common centers of positive and negative charges, and thus no dipole moment is set up without any other electric field. However if an atom or a nonpolar molecule is placed in an external electric field, the field distorts the electron orbits and separates the centers of positive and negative charges.

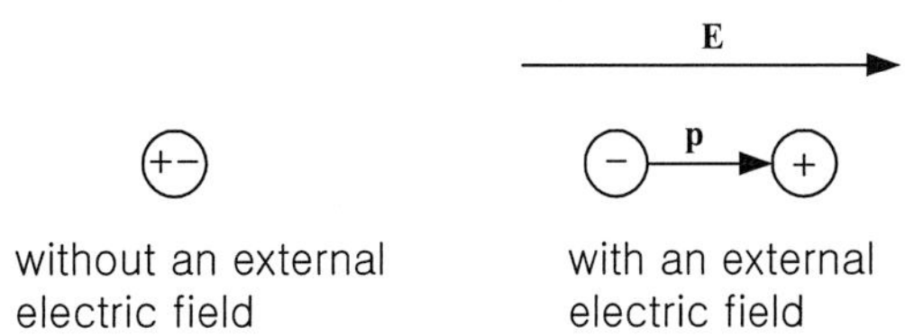

Fig. 7.24 Electrical dipole induced by the external field.

The shift sets up a dipole moment $\boldsymbol{p}$, which is said to be "induced". The atom or molecule is then said to be "polarized" by the field.

○ Calculating the electric field from the potential

$$\boldsymbol{E} = -\nabla V \tag{7.17}$$

※ Proof

From the definition of V, i.e.

$$V = \frac{U}{q_0} = -\int_{\infty}^{r} \boldsymbol{E} \cdot d\boldsymbol{r},$$

we have (Fig. 7.25)

$$\Delta V = -\boldsymbol{E} \cdot \Delta \boldsymbol{r} = -E_r \Delta r$$

or

$$E_r = -\frac{\partial V}{\partial r}$$

The component of $\boldsymbol{E}$ in the r-direction is the negative of the rate of change of the electric potential with distance in that direction. This means

$$E_x = -\frac{\partial V}{\partial x}, \quad E_y = -\frac{\partial V}{\partial y}, \quad E_z = -\frac{\partial V}{\partial z},$$

or simply

$$\boldsymbol{E} = -\nabla V$$

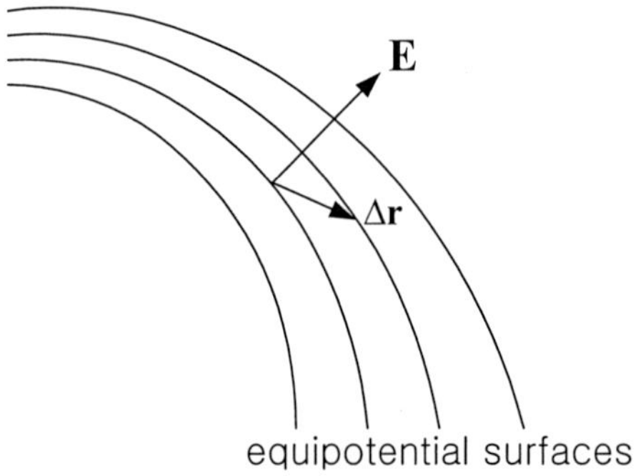

Fig. 7.25 The potential difference ΔV between a pair of points is given by the inner product of field vector $\boldsymbol{E}$ and the displacement vector $\Delta \boldsymbol{r}$ between the points.

○ Potential of an isolated charged conductor:

An excess charge placed on an isolated conductor will distribute itself on the surface of that conductor so that all points of the conductor - whether on the surface or inside - come to the same potential (Fig. 7.26)

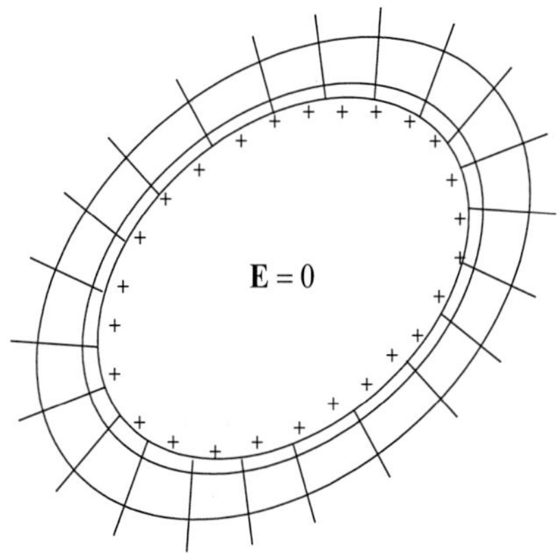

Fig. 7.26 Potential and field lines around a positively charged conductor.

○ Potential of an isolated uncharged conductor subjected to an external field (Fig. 7.27):
Field lines come into the conductor on the front face, whereas they come out from the conductor on the rear face.

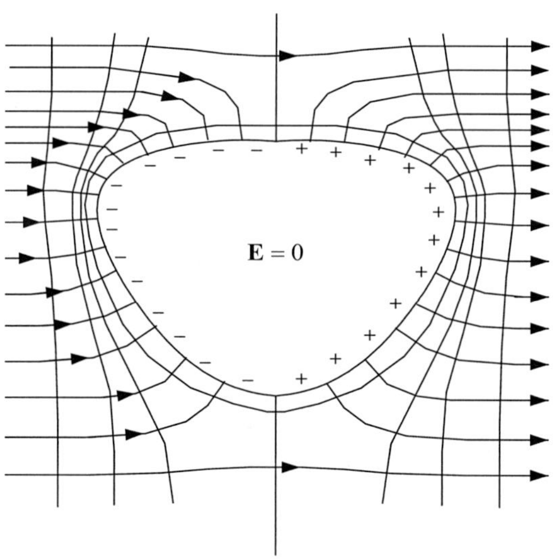

Fig. 7.27 Potential and field lines around an uncharged conductor suspended in an external electric field.

○ Potential and field around a dipole:
A dipole composed of a pair of charges $\pm q$ separated by the distance d is located

at the origin of the coordinate system (Fig. 7.28). The dipole moment vector given by this pair is $\boldsymbol{p} = qd\boldsymbol{i}$. Keeping this magnitude invariant, we can make $q \to \infty$ and $d \to 0$ simultaneously. In that case, at an arbitrary point (R, θ) the potential becomes

$$\phi = \frac{\boldsymbol{p} \cdot \hat{\boldsymbol{R}}}{4\pi\epsilon_0 R^2} = \frac{p\cos\theta}{4\pi\epsilon_0 R^2} \tag{7.18}$$

where $p = qd$. The electric field is given by

$$\boldsymbol{E} = -\nabla\phi = \left(\frac{p}{4\pi\epsilon_0}\right)\frac{2\cos\theta}{R^3}\hat{\boldsymbol{R}} + \left(\frac{p}{4\pi\epsilon_0}\right)\frac{\sin\theta}{R^3}\hat{\boldsymbol{\theta}} \tag{7.19}$$

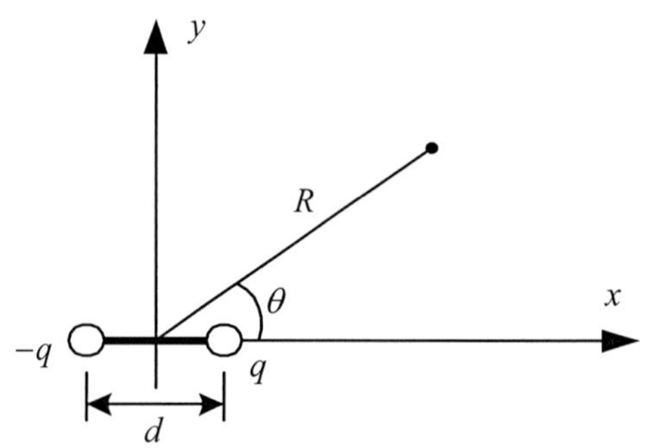

Fig. 7.28 Spherical coordinates for the formulation of the axi-symmetric field around a dipole.

7.7 Capacitance

○ Parallel-plate capacitor (Fig. 7.29)

Gauss's law states

$$q = \epsilon_0 EA = \epsilon_0 \frac{A}{d} V \tag{7.20}$$

We let $C = \epsilon_0 A/d$, the capacitance for the parallel-plate capacitor, which is a property of the geometry.

SI unit: farad = 1 coulomb per volt.

1 [F] = 1 [C/V]

1 [μF] = 10^{-6} [F]

1 [pF] = 10^{-12} [F]

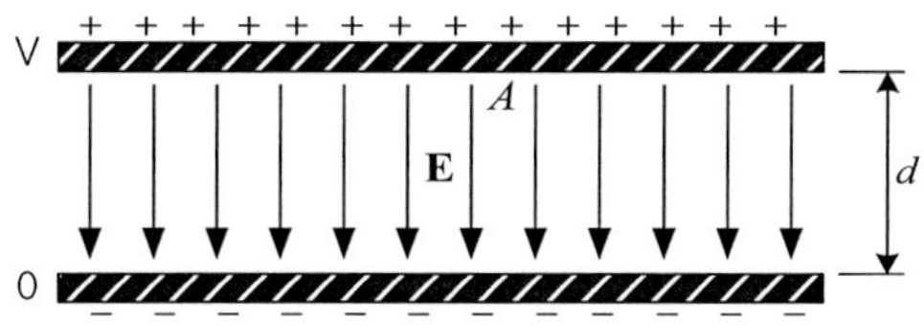

Fig. 7.29 Field vectors and charge distribution in a parallel capacitor.

○ Charging a capacitor

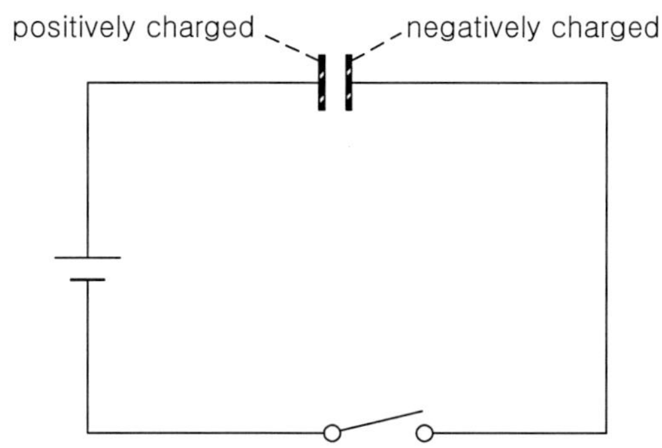

Fig. 7.30 Plates are charged by turning on the switch, and the charged state is maintained for a while after the switch is turned off.

○ Spherical capacitor

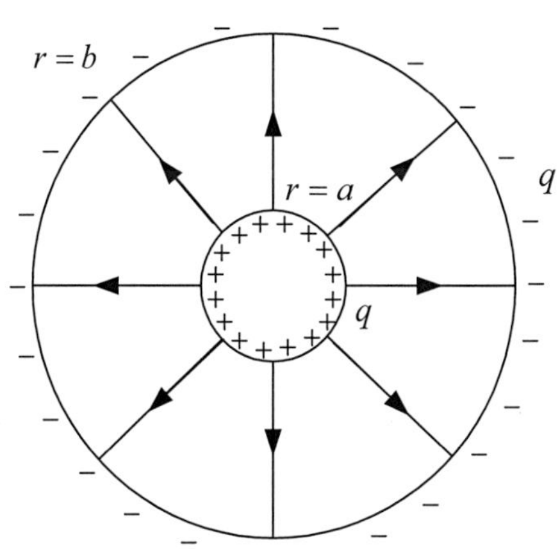

Fig. 7.31 Field lines and charge distribution in a spherical capacitor.

$$q = \epsilon_0 EA = \epsilon_0 E(4\pi r^2)$$

$$E = \frac{1}{4\pi\epsilon_0}\frac{q}{r^2}$$

$$V = \int_b^a E\,\mathrm{d}r = \frac{q}{4\pi\epsilon_0}\int_b^a \frac{\mathrm{d}r}{r^2} = \frac{q}{4\pi\epsilon_0}\frac{b-a}{ab}$$

Comparing this with $q = CV$, we get

$$C = 4\pi\epsilon_0 \frac{a}{1-a/b} \tag{7.21}$$

When $b \rightarrow \infty$ (isolated sphere of radius a),

$$C = 4\pi\epsilon_0 a \tag{7.22}$$

○ Energy stored in an electric field

Suppose that one plate is stored with the charge q'. If an extra increment of charge $\Delta q'$ is to be transferred, the work required will be (from $V' = q'/C$)

$$\Delta W = V'\Delta q' = \frac{q'}{C}\Delta q'$$

Thus, the total work required to bring the total capacitor charge up to a final value q is

$$W = \int \mathrm{d}W = \frac{1}{C}\int_0^q q'\mathrm{d}q' = \frac{q^2}{2C}$$

$$\therefore\ U = \frac{q^2}{2C} \quad \text{or} \quad U = \frac{CV^2}{2} \quad \text{(potential energy)} \tag{7.23}$$

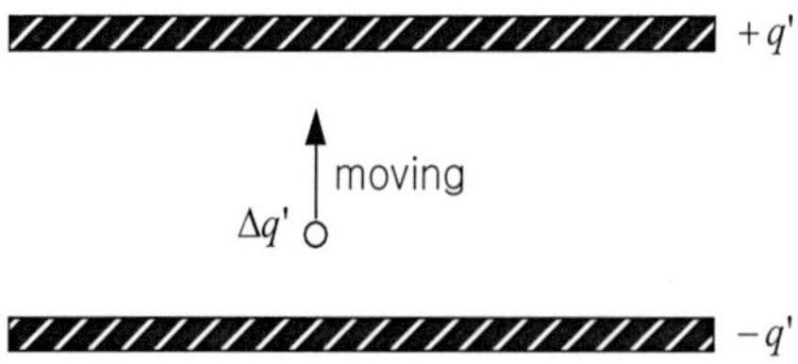

Fig. 7.32 A small test charge $\Delta q'$ moves from the negatively charged bottom plate to the positively charged top plate leading to an increase of energy.

○ Capacitor with a dielectric

A nonconducting material, such as air, glass, paper or wood, is called *dielectric.* When the space between the two conductors of a capacitor is occupied by a dielectric, the capacitance is increased by a factor κ that is characteristic of the material.

$$C = \kappa C_0 \tag{7.24}$$

where C_0 is the capacitance with a vacuum (or air with good approximation). The permittivity of the dielectric material is

$$\epsilon = \kappa\epsilon_0 \tag{7.25}$$

For a parallel-plate capacitor,

$$C = \frac{\kappa \epsilon_0 A}{d} = \frac{\epsilon A}{d} \tag{7.26}$$

The followings are data of the dielectric constant κ for some materials.

material	dielectric constant
air	1.00059
bakelite	4.9
glass	5.6
paper	3.7
plexiglass	3.4
polystyrene	2.55

○ In a region completely filled by a dielectric material of dielectric constant κ, all electrostatic equations containing the permittivity constant ϵ_0 are to be modified by replacing ϵ_0 by $\epsilon = \kappa\epsilon_0$. So, for a fixed distribution of charges, the effect of a dielectric is to weaken the electric field.

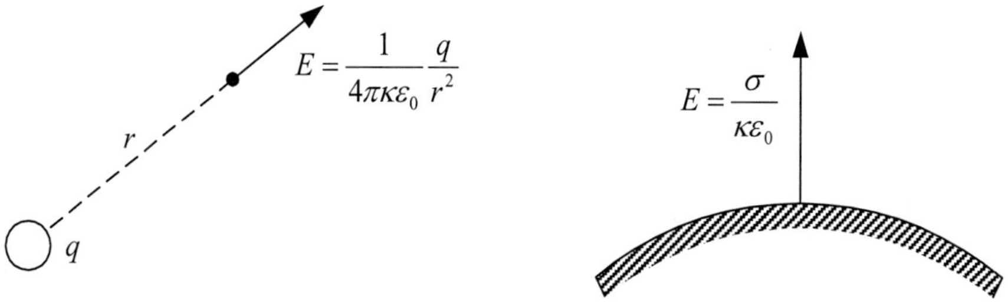

Fig. 7.33 Electric field from a point charge (left) and the field generated by a surface charge density (right) in the presence of dielectric material.

○ Dielectrics; an atomic view

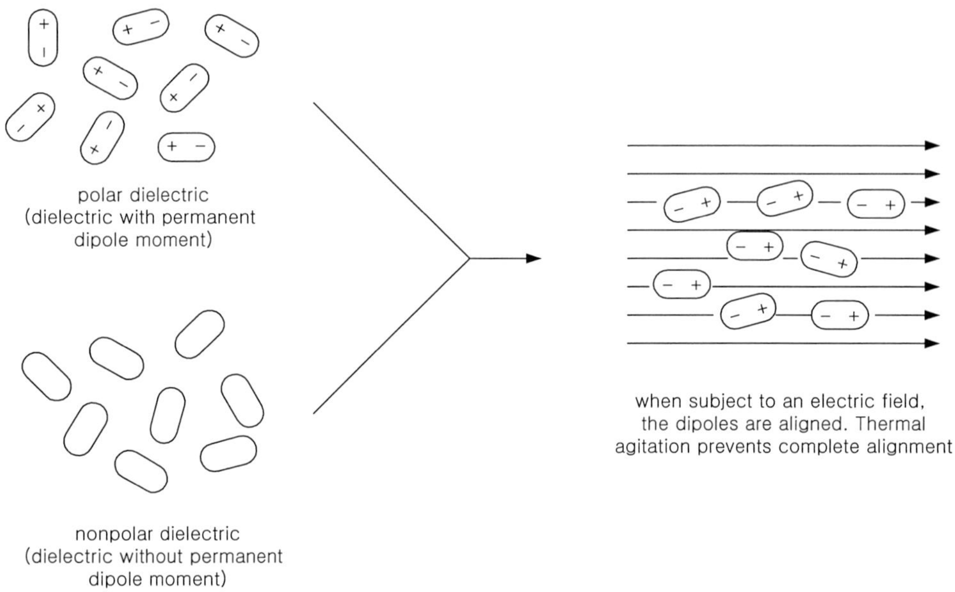

Fig. 7.34 Molecules of polar dielectric or non-polar dielectric materials tend to align with the external field.

※ Polarization of nonpolar molecule

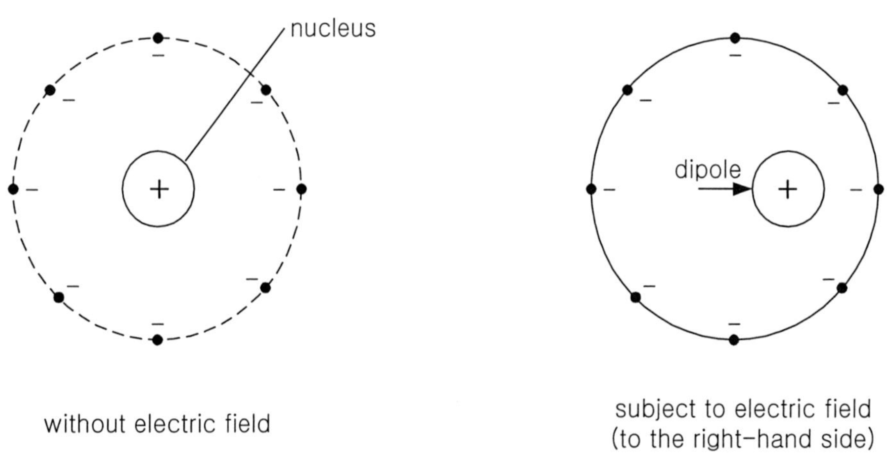

Fig. 7.35 Molecular structure without (left) and with (right) an electric field.

We have seen previously, from just an equation, that the effect of a dielectric is to weaken the electric field. Here is the explanation.

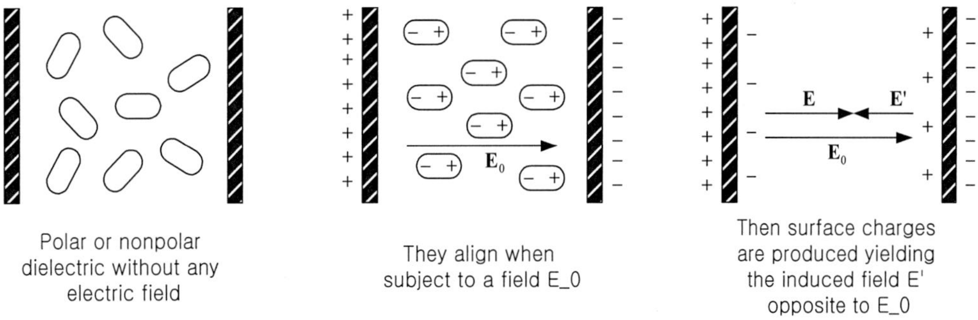

Fig. 7.36 Mechanism of field reduction in the presence of a dielectric material.

The resultant field $\boldsymbol{E}$ inside the dielectric is therefore

$$\boldsymbol{E}=\boldsymbol{E}_0+\boldsymbol{E}' \tag{7.27}$$

and its magnitude is decreased.

○ Gauss' law in dielectrics:

$$q=\epsilon_0\oint \boldsymbol{E}\cdot \mathrm{d}\boldsymbol{A} \quad \text{(Gauss' law in air or vacuum)} \tag{7.28}$$

$$q=\epsilon_0\oint \kappa\boldsymbol{E}\cdot \mathrm{d}\boldsymbol{A} \quad \text{(Gauss' law in dielectrics)} \tag{7.29}$$

Homework

1. Show that the electric field E is given by

$$E=\frac{1}{2\pi\epsilon_0}\frac{p}{z^3} \tag{7.6}$$

when two charges with q and $-q$ are very close to each other. Also prove that $p=qd$.

2. Suppose two charges are located on the walls as shown below. Determine the force acting on a particle of 1C located at O, P_1 and P_2, respectively.

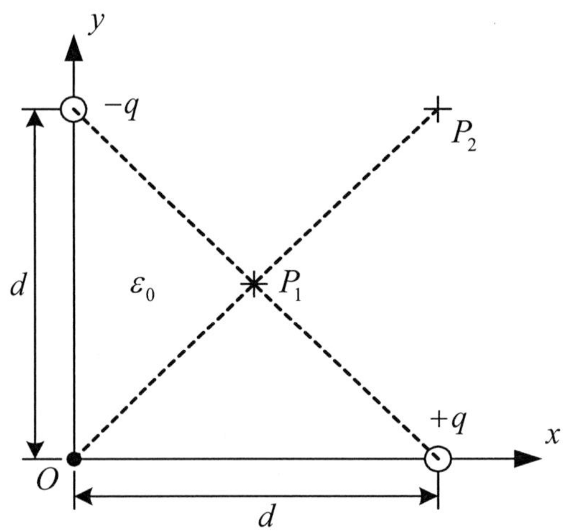

Fig. 7.37 A pair of charges located on the walls.

3. A sphere of radius a has a charge density that varies with distance r from the center according to $\rho = A\sqrt{r}$, where $A = \text{const}$. Find E in the region $r \leq a$ and in the region $r \geq a$, respectively, as a function of r.
4. Charge is distributed with a constant charge density σ on a circular disc of radius a lying in the xy plane with center at the origin. Show that the potential at a point on the z-axis is given by

$$\phi = \frac{\sigma}{2\epsilon_o}\left[(a^2 + z^2)^{1/2} - |z|\right]$$

5. Calculate the capacitance of a capacitor made of two parallel square plates, $5\,\text{cm} \times 5\,\text{cm}$ in size, separated by the distance $1\,\text{mm}$ filled with glass and air, respectively.

7.8 Current and Resistance

○ Moving charges and electric currents

$$i = \frac{dq}{dt} \quad \text{(flow rate of charges through a passage)} \tag{7.30}$$

unit = ampere; 1 [A] = 1 [C/s]

○ Current density (or current flux):

$$J = i/A \quad \text{(current per unit area)} \tag{7.31}$$

$$i = \int \boldsymbol{J} \cdot \mathrm{dA} \qquad (7.32)$$

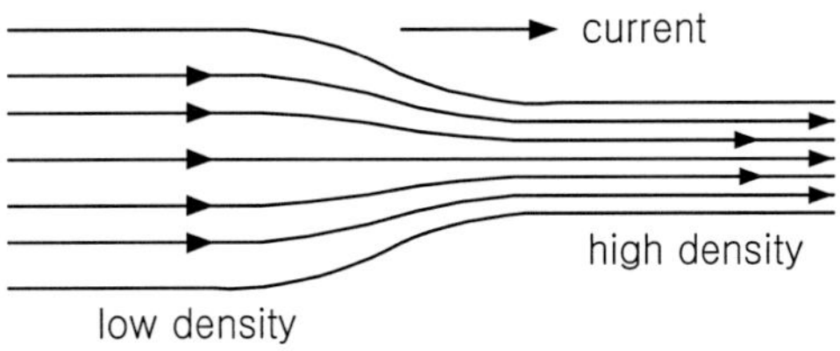

Fig. 7.38 Low and high current densities along a conductor with different cross-sectional area.

○ Resistance and resistivity

$$R = V/i \text{ (resistance): unit = ohm; } 1[\Omega] = 1[\mathrm{V/A}] \qquad (7.33)$$

The resistance is dependent not only on the material itself but on the geometry (length, area etc.) of the passage. Define resistivity (geometry-independent) as follows.

$$\rho = E/J \ [\Omega \mathrm{m}] \qquad (7.34)$$

This relationship can be written in a more rigorous way:

$$\boldsymbol{E} = \rho \boldsymbol{J} \qquad (7.35)$$

Conductivity σ is the reciprocal of the resistivity:

$$\sigma = 1/\rho \qquad (7.36)$$

Resistivity of some materials at the temperature 20°C:

Material	ρ [Ωm]
silver	1.62×10^{-8}
copper	1.69×10^{-8}
Al	2.75×10^{-8}
iron	9.68×10^{-8}
silicon, n-type	8.7×10^{-4}
glass	$10^{10} \sim 10^{14}$

If we know the resistivity of a substance, we can calculate the resistance of a length of wire made of that substance.

$$E = V/L, \quad J = i/A$$

$$\rho = E/J = \frac{V/L}{i/A} = \left(\frac{A}{L}\right)\frac{V}{i}$$

$$\frac{V}{i} = \frac{\rho L}{A} \equiv R \tag{7.37}$$

○ Ohm's law: the current through a device is always directly proportional to the potential difference applied to the device:

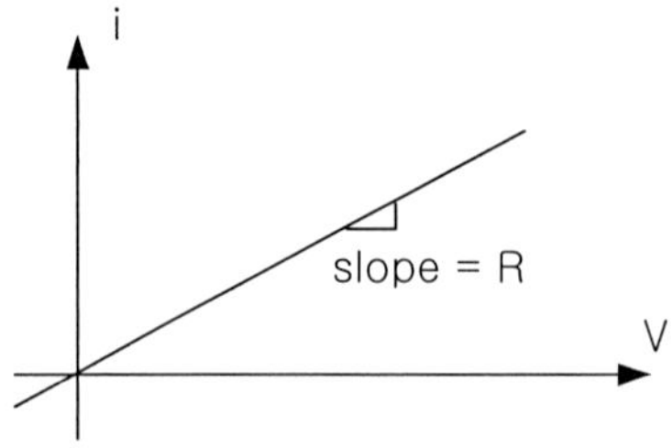

Fig. 7.39 Linear current-voltage relation, $V = Ri$, following the Ohm's law.

If we focuss on conducting materials rather than on conducting devices, we use the relation

$$E = \rho J \tag{7.38}$$

and define the Ohm's law as the case when the resistivity of the *material* is independent of the magnitude and direction (polarity) of the applied electric field.

○ Microscopic view of Ohm's law

Free electron's path during 6 collisions.

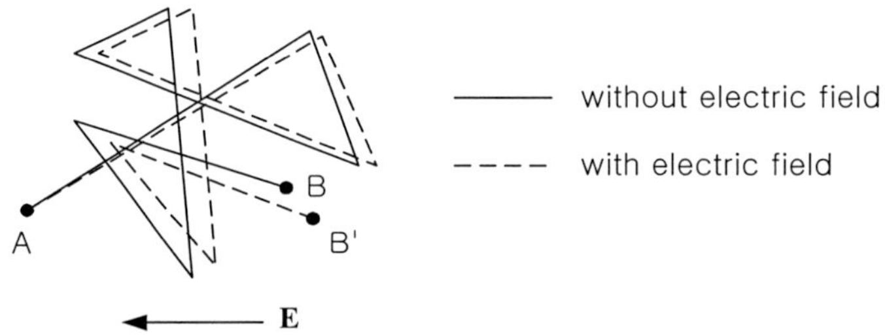

Fig. 7.40 Microscopic view of the free electron's motion with (dashed lines) and without (solid lines) the electric field.

Free electron's free speed: $v_{\mathrm{eff}} \cong 1.6 \times 10^6$ [m/s]

Drift speed with non-zero $\boldsymbol{E}$: typically, 5×10^{-7} [m/s]

The drift velocity is given by

$$v_d = a\tau$$

where a is the acceleration due to the electric field;

$$a=\frac{F}{m}=\frac{eE}{m}$$

and τ is the average time between collisions (mean free time). Then

$$v_d=\frac{eE\tau}{m}$$

On the other hand, the current flux is given from

$$J=(ne)v_d$$

where n is the number of carriers (free electrons) per unit volume. Substituting for v_d, we derive

$$E=\frac{m}{e^2 n\tau}J$$

Then we obtain the following formula for the resistivity

$$\rho=\frac{m}{e^2 n\tau} \tag{7.39}$$

○ Power in electric circuit

$$P=i\,V=i^2R=\frac{V^2}{R} \tag{7.40}$$

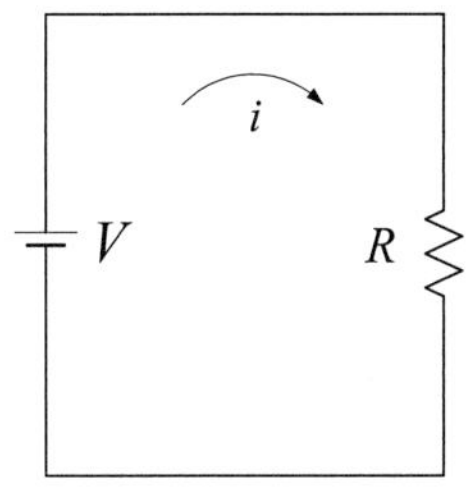

Fig. 7.41 Simplest electric circuit having one resistor.

○ Semiconductors

Pure silicon has a high resistivity. Thus it is effectively an insulator. However its resistivity can be greatly reduced in a controlled way by adding minute amount of specific "impurity" atoms in a process called "doping".

pure silicon; $\rho=2.5\times10^3$

silicon, n-type; $\rho=8.7\times10^{-4}$

silicon, p-type; $\rho=2.8\times10^{-3}$

conductor: easy to free and move the conduction electrons

insulator: hard to free and move the conduction electrons
semiconductor: it is like an insulator, but the energy required to free some electrons is not quite large.

○ Superconductor

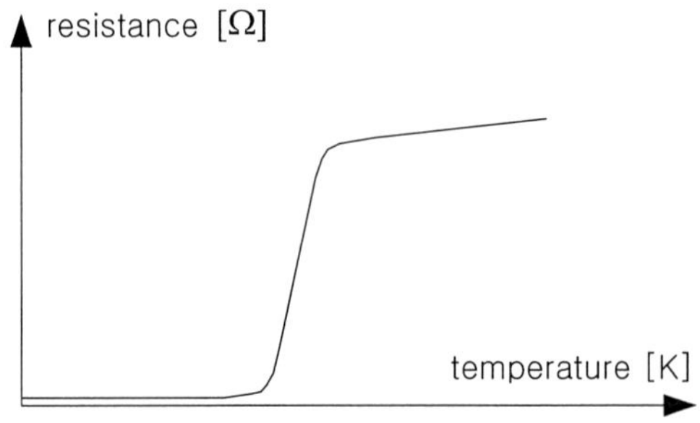

Fig. 7.42 Resistance of a typical superconductor as a function of temperature showing very low resistance at low temperatures.

7.9 Circuits

$$E - iR - \frac{q}{C} = 0$$

$$i = \frac{dq}{dt}$$

$$\therefore\ R\frac{dq}{dt} + \frac{q}{C} = E \tag{7.41}$$

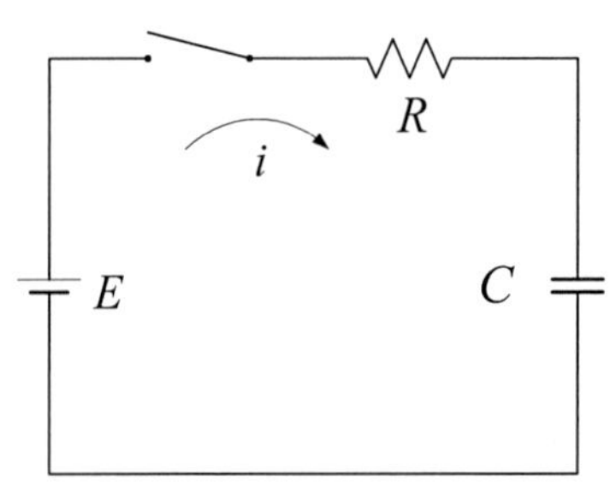

Fig. 7.43 Transient current in an R-C circuit.

7.10 Poisson Equation for the Potential

Assume the charges are continuously distributed with the volume density ρ, then the total charge within the domain will be

$$Q = \int_V \rho \, d\tau \tag{7.42}$$

where τ represents the volume for the integration. The Gauss' theorem becomes in this case

$$\oint \boldsymbol{E} \cdot d\boldsymbol{a} = \frac{Q}{\epsilon_0} = \frac{1}{\epsilon_0} \int_V \rho \, d\tau$$

Apply the divergence theorem:

$$\int_V \nabla \cdot \boldsymbol{E} \, d\tau = \frac{1}{\epsilon_0} \int_V \rho \, d\tau$$

Since this result holds for any arbitrary volume V, it is true for an infinitesimal one. Therefore

$$\nabla \cdot \boldsymbol{E} = \frac{\rho}{\epsilon_0}$$

On the other hand, $\boldsymbol{E}$ is related to the potential ϕ (in this case not V) by

$$\boldsymbol{E} = -\nabla\phi$$

Therefore

$$\nabla^2 \phi = -\frac{\rho}{\epsilon_0} \tag{7.43}$$

7.11 Dielectric Material and Polarization

As mentioned previously, when a material is subjected to an electric field, the molecules or atoms composing the material are polarized to produce dipole moments, which in turn modify the electric field inside the material. Below, we define the dipole moment and explain how the electric potential is obtained from the distribution of the dipole moments.

Suppose we have a pair of charges q and $-q$ separated by a distance d. Then the magnitude of the dipole moment is defined as $p = qd$ and its direction pointing from the point of charge q to that of charge $-q$. The electric dipole is in fact defined as the case when q is made very large and d very small but its product qd remains constant with a certain non-zero value of interest. The electric potential at a point $\boldsymbol{r}$ caused by a dipole moment $\boldsymbol{p}$ located at $\boldsymbol{r}'$ (Fig. 7.44) can be obtained from

$$\phi = \frac{\boldsymbol{p} \cdot \hat{\boldsymbol{R}}}{4\pi\epsilon_0 R^2} \tag{7.44}$$

where R is the distance measured from the dipole position to the point r, and $\hat{R}$ is the unit vector directing from the former to the latter.

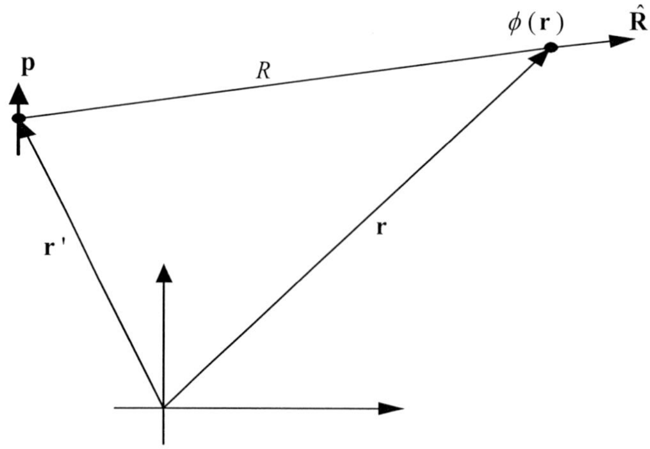

Fig. 7.44 Electric potential at a point r caused by a dipole moment p located at r'.

For a material of macroscopic scale composed of so many dipoles, their contribution to the electrostatic phenomena must be considered altogether. First we note that the dipole moment is an extensive property (property proportional to the material size such as mass or volume); for instance the momentum is an extensive property whereas the velocity (momentum per unit mass) is an intensive property. We here introduce the corresponding intensive property called 'polarization' $\boldsymbol{P}$ defined as the dipole moment per unit volume, or

$$\boldsymbol{P}=\frac{\mathrm{d}\boldsymbol{p}}{\mathrm{d}\tau} \tag{7.45}$$

where τ denotes the volume of interest. Suppose we have a confined material of volume V. When an external electric field is incident to this material, the material creates dipole moments. Let the modified electric field inside the material be $\boldsymbol{E}$, then for the linear isotropic dieletric materials the polarization $\boldsymbol{P}$ is known to be proportional to $\boldsymbol{E}$:

$$\boldsymbol{P}=\chi\epsilon_0\boldsymbol{E} \tag{7.46}$$

where

$$\chi=\kappa-1 \tag{7.47}$$

is the electric susceptibility and

$$\kappa=\epsilon/\epsilon_0 \tag{7.48}$$

is dielectric constant or relative permittivity. The polarization then induces two quantities; one is the bound volume-charge density, ρ_b, and the other the bound surface-charge density, σ_b. They are given by

$$\rho_b = -\nabla \cdot \boldsymbol{P} \tag{7.49}$$

$$\sigma_b = \boldsymbol{P} \cdot \boldsymbol{n} = P_n \tag{7.50}$$

where $\boldsymbol{n}$ indicates the unit vector normal to the surface. In Fig. 7.45, we show three cases of the electric field distribution within a material and signs of the corresponding bound volume-charge and surface-charge densities.

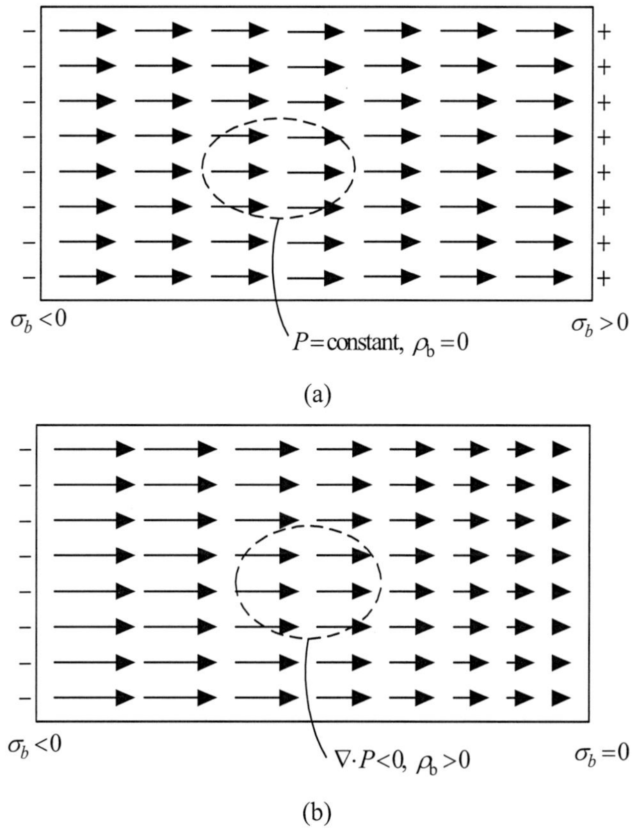

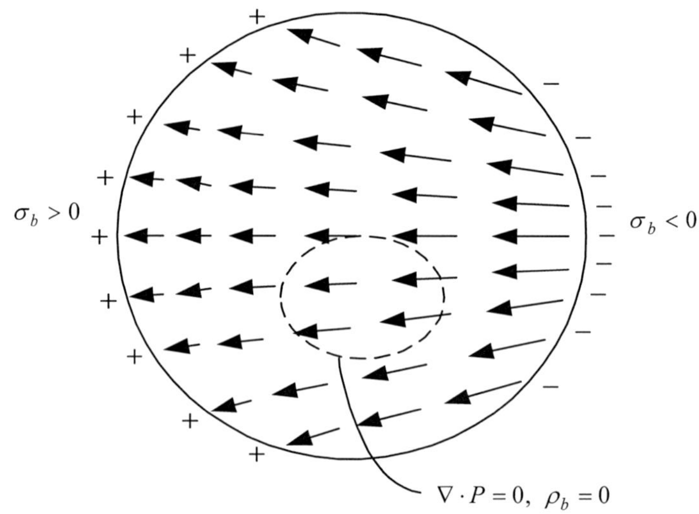

(c)

Fig. 7.45 Fundamental types in the distribution of polarization vector $\boldsymbol{P}$ and the corresponding bound volume-charge density ρ_b and bound surface-charge density σ_b: (a) uniform distribution of $\boldsymbol{P}$ --> $\rho_b = 0$, $\sigma_b < 0$ on the LHS surface and $\sigma_b > 0$ on the RHS surface; (b) linearly decreasing $\boldsymbol{P}$ --> $\rho_b > 0$, $\sigma_b < 0$ on the LHS surface and $\sigma_b = 0$ on the RHS surface; (c) non-uniform distribution of $\boldsymbol{P}$ --> possibly $\rho_b = 0$, $\sigma_b > 0$ on the LHS surface and $\sigma_b < 0$ on the RHS surface.

When two different materials are in contact with each other subject to a common external electric field (Fig. 7.46), the net bound surface-charge density is given by

$$\sigma_{b,\text{net}} = P_{1n} - P_{2n} \tag{7.51}$$

where

$$P_{1n} = \boldsymbol{P}_1 \cdot \boldsymbol{n}_1 = \boldsymbol{P}_1 \cdot \boldsymbol{n}$$

$$P_{2n} = \boldsymbol{P}_2 \cdot \boldsymbol{n}_2 = -\boldsymbol{P}_1 \cdot \boldsymbol{n}$$

and $\boldsymbol{n}$ is the unit normal vector common to both materials, $\boldsymbol{n} = \boldsymbol{n}_1 = -\boldsymbol{n}_2$.

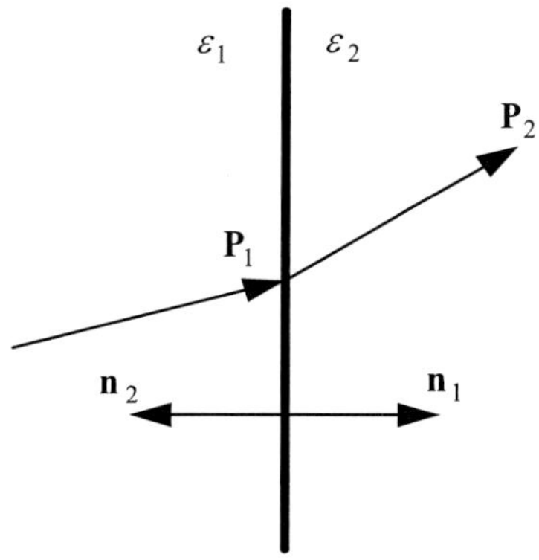

Fig. 7.46 Polarization, permittivity and normal vectors at the interface between two materials of different properties.

The potential induced by the bound charge densities is then given from

$$\phi(\boldsymbol{r}) = \frac{1}{4\pi\epsilon_0}\int_V \frac{\rho_b \mathrm{d}\tau}{R} + \frac{1}{4\pi\epsilon_0}\int_S \frac{\sigma_b \mathrm{da}}{R} \tag{7.52}$$

where $\mathrm{d}a$ denotes an infinitesimal area of the surrounding surface S (Fig. 7.47).

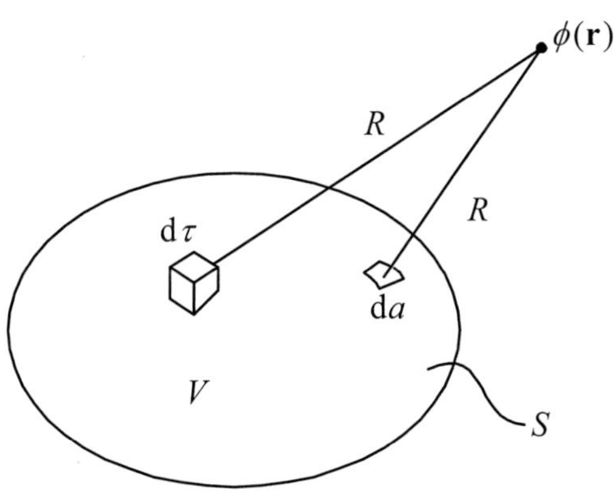

Fig. 7.47 A material volume V and its surrounding surface S, for which the potential ϕ at $\boldsymbol{r}$ is to be obtained.

As a matter of fact, the effect of the dielectric material on the electric field (or potential) can be simply understood in terms of the equivalent problem. For instance when a isolated dielectric material of permittivity ϵ is submerged in a vacuum subjected to a certain field, its equivalent problem is constructed by taking out the material but considering the distribution of bound volume-charge density ρ_b *inside* the interface and distribution of bound surface-charge density σ_b *at* the interface (Fig. 7.48). All the region is of course considered to be occupied by the vacuum.

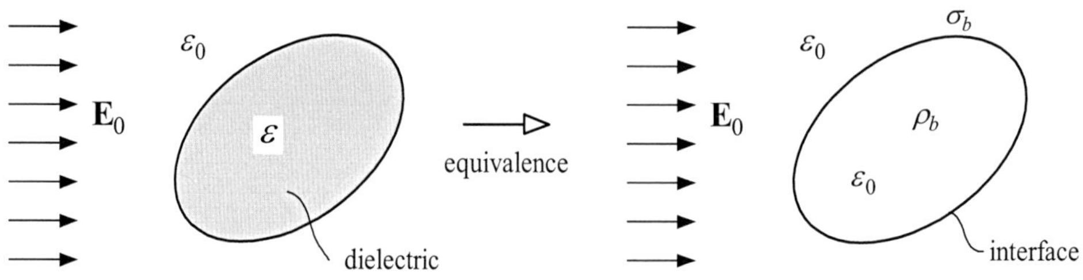

Fig. 7.48 Original configuration (left) and its equivalent model (right) of an electrostatic problem under an external field. Notice the replacement of the dielectric material in the original problem by a vacuum (or air) but with bound volume-charge and surface-charge density in the equivalent model.

As an example, we consider a capacitor with dielectric between a pair of flat plates separated by a distance d and charged by Q and $-Q$ (Fig. 7.49). First we obtain the electric field in the absence of the dielectric,

$$E_0 = \frac{\sigma_f}{\epsilon_0} \tag{7.53}$$

where $\sigma_f = Q/A$ is the free surface-charge density and A is the area of each plate. When a dielectric material is inserted between the plates, the electric field is decreased to E, which is given from $E = E_0 - E_b$. Here E_b is the electric field created by the bound surface charge density σ_b, which is the same as the polarization $P = \chi\epsilon_0 E$. Therefore

$$E_b = \frac{\chi\epsilon_0 E}{\epsilon_0} = \chi E \tag{7.54}$$

Then the equation to be solved for E is

$$E = E_0 - E_b = E_0 - \chi E$$

and we obtain

$$E = \frac{E_0}{1+\chi} = \frac{E_0}{\kappa} \tag{7.55}$$

We note that the final electric field is reduced to $1/\kappa$ times the original field; κ is always greater than 1. This means that more charges can be accommodated between the plates for the same potential difference across them, implying higher capacitance. Equivalently, this problem can be considered to be finding the electric field between a pair of plates with a *vacuum* between them but now the plates being applied not only by the free surface charge-density $\pm\sigma_f$ imposed externally but also by the induced surface-charge density $\pm\sigma_b$ which is to be determined from the final field by $\pm\chi\epsilon_0 E$.

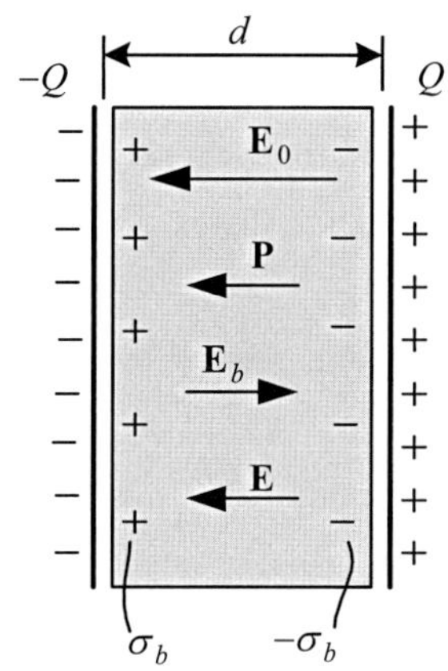

Fig. 7.49 Example problem of a parallel capacitor with a dielectric material.

7.12 Boundary Conditions for the Poisson Equation

Boundary conditions for the Poisson equation at the interface of two materials can be given in terms of the electric field components. First no jump in the field is allowed, or the field must be continuous across the interface, its equivalence being equality of the tangential components;

$$E_{1t} = E_{2t} \tag{7.56}$$

where the subscript t means the tangential component. For the normal components, the boundary condition depends on the presence of free surface charge σ_f at the interface:

$$\epsilon_2 E_{2n} - \epsilon_1 E_{1n} = \sigma_f \tag{7.57}$$

where the normal component is defined positive when the field is directing from the material 1 to the material 2. Here the surface charge density on the right-hand side of the above equation includes only the free surface-charge; that is the bound surface-charge density should not be included. If the material 1 is a dielectric with permittivity ϵ and the material 2 is vacuum, we can write this as $\epsilon_0 E_{2n} - \epsilon E_{1n} = \sigma_f$ or

$$\epsilon_0 E_{2n} - \epsilon_0 E_{1n} = \sigma_f + \sigma_b \tag{7.58}$$

where the last terms comes from the replacement of $\chi\epsilon_0 E_{1n}$ by σ_b. We can use either one of the two equations for the boundary conditions for the normal component of the electric field.

7.13 Solutions of the Simple Electrostatic Problems

In general, the electrostatic problems must be solved numerically or at least using semi-analytical methods (such as the series-solution method). But, for the problems of symmetry, we can use mathematical tools such as the method of image and the method of separation of variables to be introduced here.

7.13.1 Point charge and an infinite plane grounded conductor

Plane is located at $x=0$ and a point charge of q at $(d,0,0)$ (Fig. 7.50). Potential on the plane is set at zero. We can use the method of image (or other kinds) to get the solution

$$\phi=\frac{1}{4\pi\epsilon_0}\left(\frac{q}{R}+\frac{-q}{R'}\right) \tag{7.59}$$

where $R^2=[(x-d)^2+y^2+z^2]$ and $R'^2=[(x+d)^2+y^2+z^2]$. The electric field is given as

$$\boldsymbol{E}=-\nabla\phi=\frac{q}{4\pi\epsilon_0}\left\{\frac{[(x-d),y,z]}{R^3}-\frac{[(x+d),y,z]}{R'^3}\right\} \tag{7.60}$$

The first term corresponds to the potential to be obtained in the absence of the plane (to be referred as $\boldsymbol{E}_{\text{free}}$) whereas the second to the one induced by the presence of the plane (or the image charge) to be referred as $\boldsymbol{E}_{\text{ind}}$; in fact, the second term comes from the surface charge on the plane induced by original charge q as will be shown.

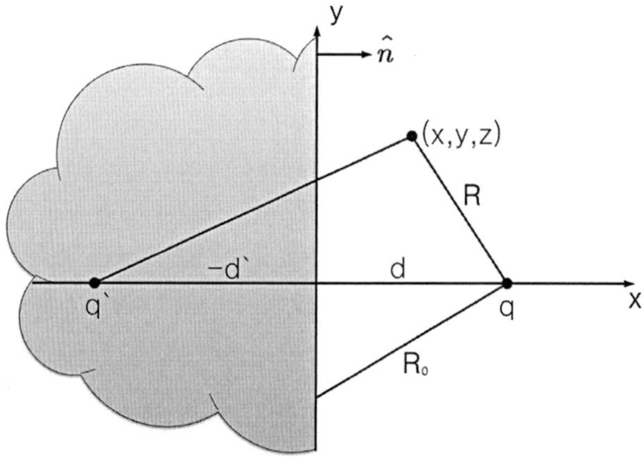

Fig. 7.50 A point charge located in air (or vacuum) at the distance d from the surface of a conductor grounded.

We note that on the plane the field vector is normal to the surface (Fig. 7.51), i.e., $E_y = E_z = 0$. On the other hand, the normal component of the field E_n is non-zero:

$$E_n = E_x = -\frac{qd}{2\pi\epsilon_0 R_0^3} \tag{7.61}$$

Therefore we expect a non-uniform distribution of surface charge density on the plane to be given from

$$\sigma = \epsilon_0 E_n = -\frac{qd}{2\pi R_0^3} \tag{7.62}$$

This surface charge is said to be *induced* by the point charge. Integrating this over the whole plane surface gives the total charge which is equal to the charge of the image, $-q$. Force acting on the point charge can be obtained from the formula $\boldsymbol{F} = q\boldsymbol{E}_{\mathrm{ind}}$ (note that $\boldsymbol{E}_{\mathrm{free}}$ does not contribute to the force).

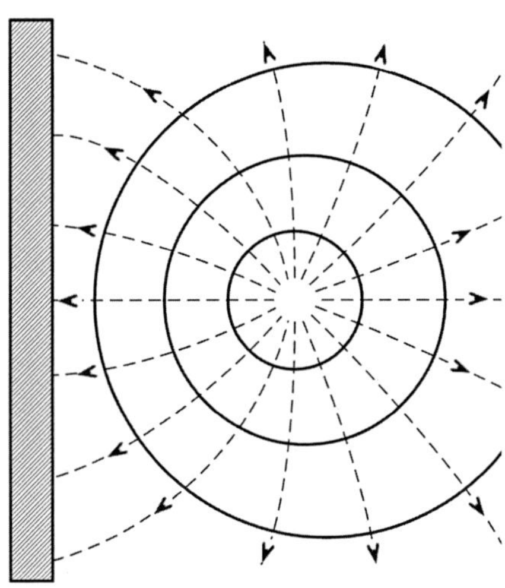

Fig. 7.51 Equi-potential lines (solid) and electric field lines (dashed) for the problem of Fig. 7.50.

7.13.2 Point charge and a grounded conducting sphere

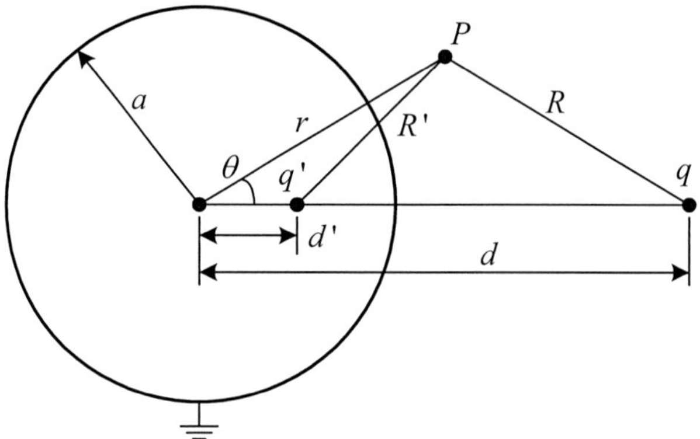

Fig. 7.52 A point charge located in air at the distance d from the center of a spherical conductor grounded.

A grounded conducting sphere is located at the coordinate origin and a point charge of q at $(d, 0, 0)$ as shown in Fig. 7.52. Potential on the planes is set at zero. The potential can be obtained by introducing another charge q' within the sphere.

$$\phi = \frac{1}{4\pi\epsilon_0}\left(\frac{q}{R} + \frac{q'}{R'}\right) \tag{7.63}$$

where the charge q' and its distance from the origin, d', are to be determined from the condition of zero potential on the sphere leading to $q' = -aq/d$ and $d' = a^2/d$. The surface charge is also induced in this case, and since the sphere is grounded, the total amount of the surface charge should be non-zero and equal to q'. The original point charge q is attracted toward the sphere and its magnitude can be given by the Coulomb force between q and its image q'.

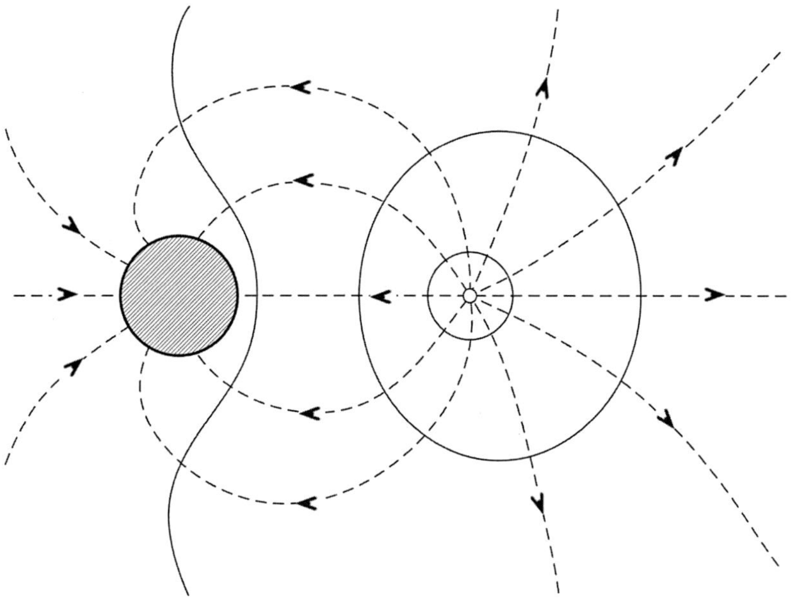

Fig. 7.53 Equi-potential lines (solid) and electric field lines (dashed) for the problem of Fig. 7.52.

7.13.3 Point charge and an insulated uncharged conducting sphere

This is a variation of the previous example. We now assume that the conducting sphere is isolated (insulated) and initially neutral (Fig. 7.54). This means that the potential remains again at a constant value on the sphere but the total amount of the sphere's surface charge must vanish. This can be achieved just introducing another point charge $-q'$ at the center of the sphere in addition to the two point charges presented in the previous example; sum of the two point charges within the sphere makes zero (so the total amount of the surface charge is zero) and the equi-potential lines of the potential caused by the added single point charge are spheres (so the surface potential remains uniform).

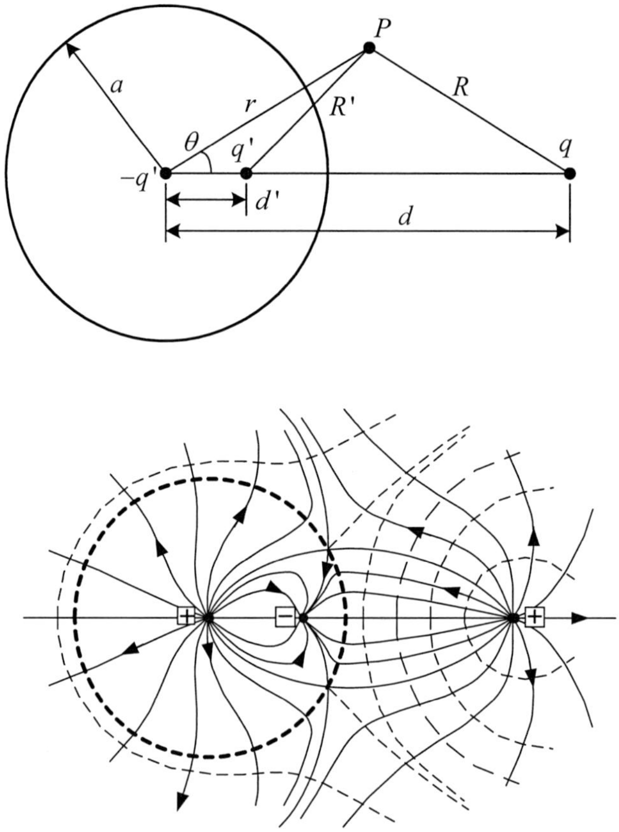

Fig. 7.54 A point charge located in air at the distance d from the center of an isolated (insulated) spherical conductor (top). Equi-potential lines (solid) and electric field lines (dashed) obtained for this problem (bottom).

7.13.4 Point charge and an infinite dielectric plane

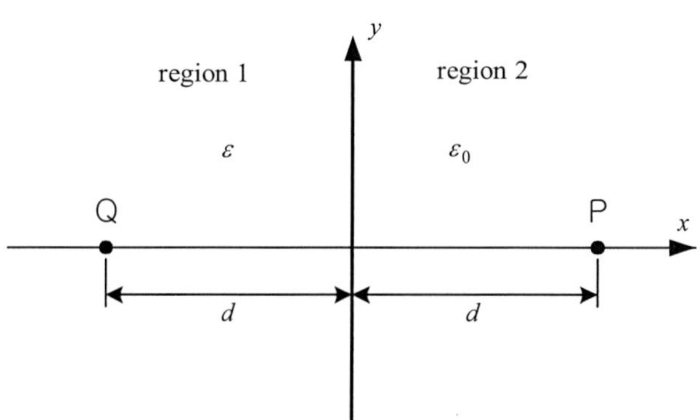

Fig. 7.55 A point charge located in air at the distance d from the surface of a dielectric solid.

This is a variant of the first example (Fig. 7.50), and the planar conductor here is replaced by a planar dielectric material of the permittivity ϵ (Fig. 7.55). The potential in the vacuum (region 2), ϕ_2 can be written as the sum of the potential caused by the charge q located at the point P in the region 2 and the image q' located at the point Q in the region 1;

$$\phi_2 = \frac{1}{4\pi\epsilon_0}\left(\frac{q}{R} + \frac{q'}{R'}\right) \tag{7.64}$$

where R and R' are defined in Fig. 7.50. The unknown charge q' is to be determined shortly from the application of the boundary conditions. The corresponding electric field vector is

$$\boldsymbol{E}_2 = -\nabla\phi_2 = \frac{1}{4\pi\epsilon_0}\left\{\frac{q[(x-d), y, z]}{R^3} + \frac{q'[(x+d), y, z]}{R'^3}\right\} \tag{7.65}$$

On the other hand, the potential in the region 1 can be described by a single image charge q'' at the point P in the above figure:

$$\phi_1 = \frac{1}{4\pi\epsilon_0}\left(\frac{q''}{R}\right) \tag{7.66}$$

Note that the permittivity is set at ϵ_0 and not ϵ. The corresponding electric field is

$$\boldsymbol{E}_1 = -\nabla\phi_1 = \frac{1}{4\pi\epsilon_0}\left\{\frac{q''[(x-d), y, z]}{R^3}\right\} \tag{7.67}$$

Boundary conditions to be applied are:

$$\epsilon_0 E_{2x} = \epsilon E_{1x} \quad \text{at } x = 0$$
$$E_{2y} = E_{1y} \quad \text{at } x = 0$$
$$E_{2z} = E_{1z} \quad \text{at } x = 0$$

Applying these boundary conditions, we obtain the followings.

$$q' = -\left(\frac{\kappa - 1}{\kappa + 1}\right) q \tag{7.68}$$

$$q'' = \frac{2q}{\kappa + 1} \tag{7.69}$$

The case with $\kappa = 1$ (vacuum everywhere) results in $q' = 0$ and $q'' = q$. Therefore both the regions 1 and 2 take the same solution form as it should be. The limit $\kappa \to \infty$ (conduction plane) on the other hand produces $q' = -q$ and $q'' = 0$ corresponding to the case of example 1. The induced surface charge can be derived from $\epsilon_0(E_{2x} - E_{1x}) = \sigma_b$:

$$\sigma_b = \frac{-qd}{2\pi R^3}\frac{\kappa - 1}{\kappa + 1} \tag{7.70}$$

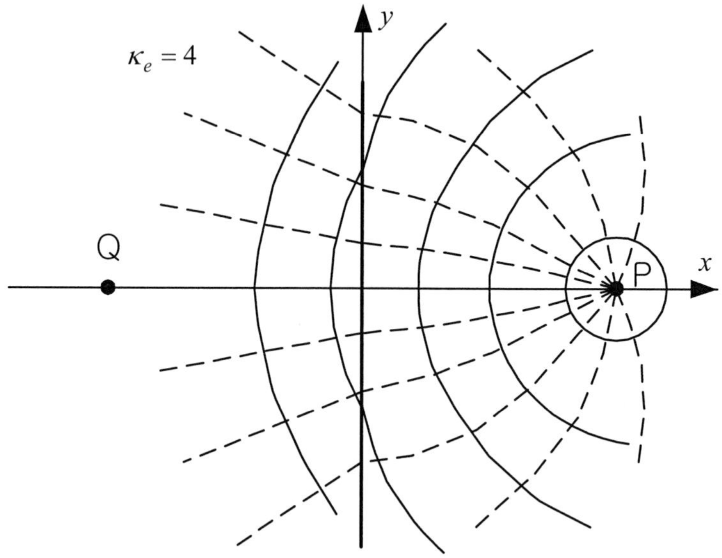

Fig. 7.56 Equi-potential lines (solid) and electric field lines (dashed) for the problem of Fig. 7.55.

7.13.5 Dielectric sphere in a uniform field

Let's put a sphere of dielectric material of permittivity ϵ with radius a in a homogeneous electric field $\boldsymbol{E}_0$ applied between a pair of parallel plates (Fig. 7.57). To solve this problem we use axi-symmetric spherical coordinates (r,θ). Then, for the axi-symmetric problem, the equation to be solved is

$$\frac{1}{r^2}\frac{\partial}{\partial r}\left(r^2\frac{\partial \phi}{\partial r}\right)+\frac{1}{r^2\sin\theta}\frac{\partial}{\partial \theta}\left(\sin\theta\frac{\partial \phi}{\partial \theta}\right)=0 \tag{7.71}$$

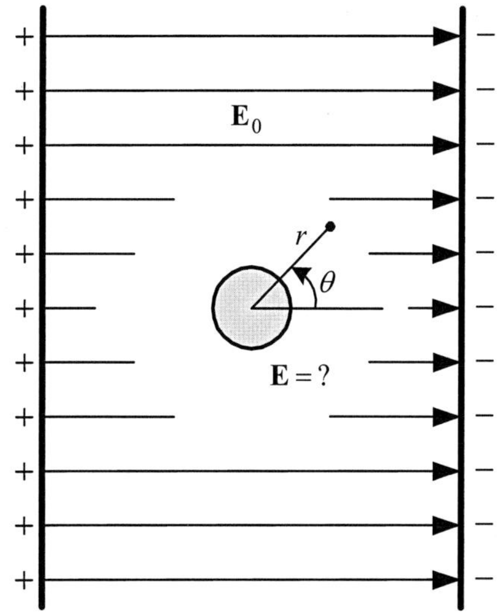

Fig. 7.57 A dielectric sphere with permittivity ϵ subjected to uniform electric field $\boldsymbol{E}_0$.

We use the method of separation of variables. For convenience, we split the solution into the one for the outer region, ϕ_o, and the one for the inner region, ϕ_i. The boundary conditions to be applied are

$$\phi_o \rightarrow -E_0 x = -E_0 r\cos\theta \text{ as } r\rightarrow\infty$$

$$\epsilon_0\left(\frac{\partial\phi_o}{\partial r}\right)_{r=a} = \epsilon\left(\frac{\partial\phi_i}{\partial r}\right)_{r=a}$$

$$\left(\frac{1}{r}\frac{\partial\phi_o}{\partial\theta}\right)_{r=a} = \left(\frac{1}{r}\frac{\partial\phi_i}{\partial\theta}\right)_{r=a}$$

For the region outside the sphere, the solution takes the form

$$\phi_o = \left(A_o r + \frac{B_o}{r^2}\right)\cos\theta \tag{7.72}$$

From the boundary condition at infinity, we require $A_0 = -E_0$. The solution for the inner region is

$$\phi_i = \left(A_i r + \frac{B_i}{r^2}\right)\cos\theta \tag{7.73}$$

We require $B_i = 0$ to avoid singularity at the origin. We are left with two unknown constants, B_o and A_i. These can be obtained by applying the remaining boundary conditions to yield

$$\phi_o = -E_0 r\cos\theta + \frac{\kappa-1}{\kappa+2}\frac{a^3 E_0 \cos\theta}{r^2} \qquad (7.74)$$

$$\phi_i = -\left(\frac{3E_0}{\kappa+2}\right) r\cos\theta \qquad (7.75)$$

The field inside the sphere is

$$\boldsymbol{E}_i = \left(\frac{3}{\kappa+2}\right)\boldsymbol{E}_0 \qquad (7.76)$$

Note that the induced inside field is parallel to but less than $\boldsymbol{E}_0$ in its magnitude. For $\kappa=1$ (vacuum inside) we have $\boldsymbol{E}_i = \boldsymbol{E}_0$, and for $\kappa\to\infty$ (conductor) it approaches zero. The polarization is computed from $\boldsymbol{P} = \chi\epsilon_0 \boldsymbol{E}_i$ or

$$\boldsymbol{P} = \left(\frac{\kappa-1}{\kappa+2}\right) 3\epsilon_0 \boldsymbol{E}_0 \qquad (7.77)$$

The magnitude of the total dipole moment is given from $p = PV$, where V denotes the volume of the sphere:

$$p = 4\pi\left(\frac{\kappa-1}{\kappa+2}\right) a^3 \epsilon_0 \boldsymbol{E}_0 \qquad (7.78)$$

Then we can write ϕ_o as

$$\phi_o = -E_0 r\cos\theta + \frac{p\cos\theta}{4\pi\epsilon_0 r^2} \qquad (7.79)$$

Here, the second term on the right-hand side corresponds to the potential created by a dipole located at the origin. From the solution of $\boldsymbol{E}_i$ we can deduce

$$\boldsymbol{E}_i = \boldsymbol{E}_0 - \frac{\boldsymbol{P}}{3\epsilon_0} \qquad (7.80)$$

implying that the electric field inside the sphere is the sum of the original field plus the local field that is proportional to the polarization and opposite to it. Fig. 7.58 presents some of the electric field lines drawn from the solutions.

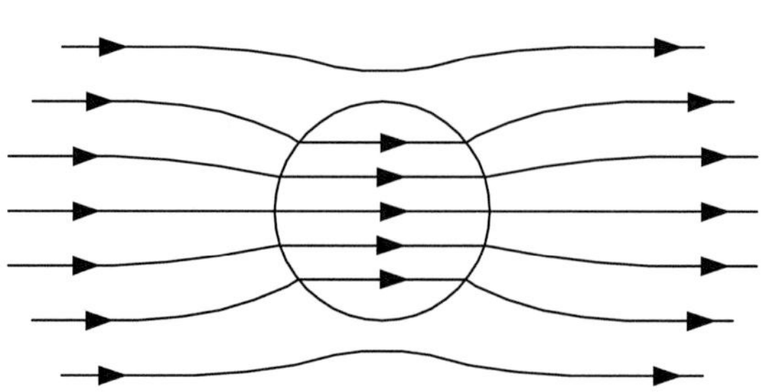

Fig. 7.58 Electric field lines for the problem of Fig. 7.57.

Homework

1. Suppose we have a pair of point charges at $(x,y,z)=(d,0,0)$ and $(x,y,z)=(-d,0,0)$, respectively, with the same a mount of charge Q (same sign). Derive the formula calculating potential as well as the field (a) on the line $y=z=0$ and (b) on the line $x=y=0$.
2. In the problem of "point charge on an infinite plane of grounded conductor", show that the sum (in fact, areal integral of the surface charge over the whole surface) of the induced surface charge on the plane is the same as the point charge.
3. In the problem of "point charge and a grounded conducting sphere", derive the expressions, (a) $q'=-aq/d$ and (b) $d'=a^2/d$ from the requirements given in the text.
4. In the problem of "point charge and insulated uncharged conducting sphere", derive the expressions for (a) q' and (b) d' to satisfy the requirements specified in the text.

| 저자 소개 |

서 용 권

서울대학교 기계공학과 졸업
State Univ. of New York at Buffalo 석사, 박사
현대자동차(주) 근무
영남공업전문대학 자동차과 근무
현재 동아대학교 기계공학과 교수

강 상 모

서울대학교 기계공학과 졸업
서울대학교 기계공학과 석사
University of Michigan(USA) 박사
대우중공업(주) 근무
대구대학교 자동차산업기계공학부 근무
현재 동아대학교 기계공학과 교수

본 교재는 2011년도 산업통상자원부의 재원으로 한국에너지기술평가원(KETEP)의 지원을 받아 수행한 연구 성과(No. 20114030200030)의 일부입니다.

정가 12,000원

다중물리 해석학 입문

2013년 6월 10일 초판 인쇄
2013년 6월 20일 초판 발행
저 자 : 서용권 · 강상모
발행자 : 우명찬 · 송 준
발행처 : 홍릉과학출판사
주 소 : 서울시 강북구 인수동 455-60
142-885
등 록 : 1976년 10월 21일 제5-66호
전 화 : (02) 999-2274~5, 903-7037
팩 스 : (02) 905-6729
e-mail: hongpub@hongpub.co.kr
http://www.hongpub.co.kr
ISBN: 978-89-97570-89-8

저자와의 협의하에 인지생략